ANALYTICAL TOOLS AND INDUSTRIAL APPLICATIONS FOR CHEMICAL PROCESSES AND POLYMERIC MATERIALS

ANALYTICAL TOOLS AND INDUSTRIAL APPLICATIONS FOR CHEMICAL PROCESSES AND POLYMERIC MATERIALS

Edited by
Slavcho Kirillov Rakovsky, DSc, Ryszard Kozlowski, PhD and Nekane Guarrotxena, PhD

Gennady E. Zaikov, DSc, and A. K. Haghi, PhD
Reviewers and Advisory Board Members

Apple Academic Press

TORONTO   NEW JERSEY

Apple Academic Press Inc. | Apple Academic Press Inc.
3333 Mistwell Crescent | 9 Spinnaker Way
Oakville, ON L6L 0A2 | Waretown, NJ 08758
Canada | USA

©2014 by Apple Academic Press, Inc.

First issued in paperback 2021

Exclusive worldwide distribution by CRC Press, a member of Taylor & Francis Group

No claim to original U.S. Government works

ISBN 13: 978-1-77463-294-9 (pbk)
ISBN 13: 978-1-926895-66-6 (hbk)

This book contains information obtained from authentic and highly regarded sources. Reprinted material is quoted with permission and sources are indicated. Copyright for individual articles remains with the authors as indicated. A wide variety of references are listed. Reasonable efforts have been made to publish reliable data and information, but the authors, editors, and the publisher cannot assume responsibility for the validity of all materials or the consequences of their use. The authors, editors, and the publisher have attempted to trace the copyright holders of all material reproduced in this publication and apologize to copyright holders if permission to publish in this form has not been obtained. If any copyright material has not been acknowledged, please write and let us know so we may rectify in any future reprint.

Trademark Notice: Registered trademark of products or corporate names are used only for explanation and identification without intent to infringe.

Library of Congress Control Number: 2012951950

Library and Archives Canada Cataloguing in Publication

Analytical tools and industrial applications for chemical processes and polymeric materials /edited by Slavcho Kirillov Rakovsky, DSc, Ryszard Kozlowski, PhD, and Nekane Guarrotxena, PhD; Gennady E. Zaikov, DSc, and A.K. Haghi, PhD, reviewers and advisory board members.

Includes bibliographical references and index.
ISBN 978-1-926895-66-6
1. Polymers--Industrial applications. 2. Polymer engineering. I. Rakovsky, S. (Slavtcho), editor of compilation II. Kozlowski, Ryszard, editor of compilation III. Guarrotxena, Nekane, editor of compilation

TA455.P58A53 2013   620.1'92   C2013-906888-0

Apple Academic Press also publishes its books in a variety of electronic formats. Some content that appears in print may not be available in electronic format. For information about Apple Academic Press products, visit our website at **www.appleacademicpress.com** and the CRC Press website at **www.crcpress.com**

ABOUT THE EDITORS

Slavcho Kirillov Rakovsky, DSc

Slavcho Kirillov Rakovsky, DSc, is a doctor of science and Professor and Director of the Institute of Catalysis at the Bulgarian Academy of Sciences, Sofia, Bulgaria. He is world renowned scientist in the field of chemical kinetics and chemical physics, chemistry of ozone, and elastomers (rubbers). He is a contributor to 35 books and has published 500 original papers as well as reviews.

Ryszard Kozlowski, PhD

Professor Ryszard Kozlowski, PhD, is the author and co-author of more than 250 original papers, 21 patents and know-how licenses, and 24 implemented technologies. He also acted as supervisor of doctoral theses, including those of a Chinese PhD student. He leads the activity of the Institute for Engineering of Polymer Materials and Dyes, Branch House for Elastomers and Rubber Processing, Piastow, Poland. Professor Kozlowski initiated and conducted interdisciplinary research on broadening raw material resources and utilization of natural fibers (flax, hemp, wool, and silk), retting, extracting and processing of bast fibers, environmental protection, fire protection, utilization of by-products and wastes from the bast fiber plants industry, and biodeterioration and its prevention. His special interest is in the flammability and toxicity of diversified fibers, polymers, fabrics, and products, including biocomposites, with a focus on the high practical aspect of this research. His specialty is flame retardancy of polymers and modern environmentally friendly composites, and he conducts applied science in the scope of natural polymeric materials.

Nekane Guarrotxena, PhD

Nekane Guarrotxena, PhD, is a Scientist at the Institute of Polymer Science and Technology in Madrid, Spanish National Research Council, Spain. She is a well-known scientist in the field of organic chemistry, chemistry and physics of polymers, and composites and nanocomposites. She has published 10 books and volumes and 500 original papers and reviews. She was the Vice-Director of the Institute of Polymer Sciences and Technology from 2001–2005. From 2008–2011, she was a visiting professor in the Department of Chemistry, Biochemistry and Materials at the University of California, Santa Barbara, and the Center for Chemistry at the Space-Time Limit (CaS-TL) at the University of California, Irvine. She is an editorial board member of the *Polymer Research* and *ISRN* journal. Her research interests focus on the synthesis and assembly of hybrid nanomaterials, nanoplasmonics, and their uses in nanobiotechnology applications, such as bioimaging, drug delivery, therapy, and biosensing.

REVIEWERS AND ADVISORY BOARD MEMBERS

A. K. Haghi, PhD

A. K. Haghi, PhD, holds a BSc in urban and environmental engineering from the University of North Carolina (USA); a MSc in mechanical engineering from North Carolina A&T State University (USA); a DEA in applied mechanics, acoustics and materials from Université de Technologie de Compiègne (France); and a PhD in engineering sciences from Université de Franche-Comté (France). He is the author and editor of 65 books as well as 1000 published papers in various journals and conference proceedings. Dr. Haghi has received several grants, consulted for a number of major corporations, and is a frequent speaker to national and international audiences. Since 1983, he served as a professor at several universities. He is currently Editor-in-Chief of the *International Journal of Chemoinformatics and Chemical Engineering* and *Polymers Research Journal* and on the editorial boards of many international journals. He is also a faculty member of University of Guilan (Iran) and a member of the Canadian Research and Development Center of Sciences and Cultures (CRDCSC), Montreal, Quebec, Canada.

Gennady E. Zaikov, DSc

Gennady E. Zaikov, DSc, is Head of the Polymer Division at the N. M. Emanuel Institute of Biochemical Physics, Russian Academy of Sciences, Moscow, Russia, and professor at Moscow State Academy of Fine Chemical Technology, Russia, as well as professor at Kazan National Research Technological University, Kazan, Russia. He is also a prolific author, researcher, and lecturer. He has received several awards for his work, including the the Russian Federation Scholarship for Outstanding Scientists. He has been a member of many professional organizations and on the editorial boards of many international science journals.

CONTENTS

LIST OF CONTRIBUTORS

Afanasov, I.
N. N. Semenov Institute of chemical Physics, Russian Academy of Sciences, 4 Kosygin str., Moscow 119991, Russia.

Akhmetkhanov, R. M.
The Bashkir State University, Ufa, Zaki Validi str. 32.

Aleksandrov, E. N.
Emanuel Institute of Biochemical Physics, Russian Academy of Sciences, ul. Kosygina 4, Moscow, 119334 Russia, E-mail: 28en1937@mail.ru/Semenov Institute of Chemical Physics, Russian Academy of Sciences, ul. Kosygina 4, Moscow, 119334 Russia. Tel.: +7 499 137 7249; Fax: +7 495 651 2191; E-mail: icp@chph.ras.ru.

Aleksandrov, P. E.
Emanuel Institute of Biochemical Physics, Russian Academy of Sciences, ul. Kosygina 4, Moscow, 119334 Russia, E-mail: 28en1937@mail.ru / Semenov Institute of Chemical Physics, Russian Academy of Sciences, ul. Kosygina 4, Moscow, 119334 Russia. Tel.: +7 499 137 7249; Fax: +7 495 651 2191; E-mail: icp@chph.ras.ru.

Andriasyan, Yu. O.
Institute of Biochemical Physics, Russian Academy of Sciences, Moscow, Russia.

Arzamasova, T. M.
Department of Molecular pharmacology and radiobiology, Medico-Biological Faculty, N. I. Pirogov Russian National Research Medical University (RNRMU), Moscow, Russia.

Babkin, V. A.
Volgograd State Architect-build University, Sebrykov Department.

Belousova, A. L.
State University of Fine Chemical Technologies, Moscow, Russia.

Berlin, A. A.
N. N. Semenov Institute of chemical Physics, Russian Academy of Sciences, 4 Kosygin str., Moscow 119991, Russia.

Chalykh, A. E.
A. N. Frumkin Institute of Physical Chemistry and Electrochemistry .

Chukicheva, I. Yu.
The Institute of chemistry of Komi of centre of science of the Ural branch of the Russian Academy of Sciences, Syktyvkar, Pervomayskaya str., 48, Moscow, Russia.

Evteeva, N. M.
Establishment of the Russian Academy of Sciences Institute of biochemical physics of N.M. Emanuelja, 119991 Moscow, Kosygina, 4; Fax: 095 1374101.

Filatov, Yu. N.
L. Ya. Karpov Physicochemical Research Institute.

Gabitov, I. T.
The Bashkir State University, Ufa, Zaki Validi str. 32.

Garipov, R. M.
Kazan National Research Technological University.

Gerasimov, K.
A. N. Frumkin Institute of Physical Chemistry and Electrochemistry.

Haghi, A. K.
University of Guilan, Rasht, Iran, E-mail: Haghi@Guilan.ac.ir.

Iordanskii, A. L.
N. N. Semenov Institute of chemical Physics, Russian Academy of Sciences, 4 Kosygin str., Moscow 119991, Russia, Tel: +7 495 939–7434. E mail: aljordan08@gmail.com.

Khovanets, G. I.
Department of Physico-Chemistry of Combustible Minerals.

Kolesov, S. V.
The Bashkir State University, Ufa, Zaki Validi str. 32.

Konstantinova, M. L.
N. M. Emanuel Institute of Biochemical Physics, Russian Academy of Sciences, Moscow.

Kornev, A. E.
State University of Fine Chemical Technologies, Moscow, Russia.

Koroteev, A. M.
Moscow State Pedagogical University, Department of Organic Chemistry, Moscow, Russia.

Koroteev, M. P.
Moscow State Pedagogical University, Department of Organic Chemistry, Moscow, Russia.

Kuchin, A. V.
The Institute of chemistry of Komi of centre of science of the Ural branch of the Russian Academy of Sciences, Syktyvkar, Pervomayskaya str. 48.

Kuhareva, T. S.
Moscow State Pedagogical University, Department of Organic Chemistry, Moscow, Russia.

Kuzmicheva, G. M.
Lomonosov Moscow University of Fine Chemical Technology.

Kuznetsov, N. M.
Emanuel Institute of Biochemical Physics, Russian Academy of Sciences, ul. Kosygina 4, Moscow, 119334 Russia, E-mail: 28en1937@mail.ru/Semenov Institute of Chemical Physics, Russian Academy of Sciences, ul. Kosygina 4, Moscow, 119334 Russia. Tel.: +7 499 137 7249; Fax: +7 495 651 2191; E-mail: icp@chph.ras.ru.

Leonova, V. B.
N. M. Emanuel Institute of Biochemical Physics, Russian Academy of Sciences, Moscow.

Lidzhi–Goryaev, V. Yu.
Emanuel Institute of Biochemical Physics, Russian Academy of Sciences, ul. Kosygina 4, Moscow, 119334 Russia, E-mail: 28en1937@mail.ru/Semenov Institute of Chemical Physics, Russian Academy of Sciences, ul. Kosygina 4, Moscow, 119334 Russia. Tel.: +7 499 137 7249; Fax: +7 495 651 2191; E-mail: icp@chph.ras.ru.

Lomakin, S. M.
Establishment of the Russian Academy of Sciences Institute of biochemical physics of N. M.Emanuelja, 119991 Moscow, Kosygina, 4; Fax: 095 1374101.

Madyuskin, N. N.
Establishment of the Russian Academy of Sciences Institute of biochemical physics of N. M.Emanuelja, 119991 Moscow, Kosygina, 4; Fax: 095 1374101.

Matyoushin, A. I.
M. Emanuel Institute of Biochemical Physics, Russian Academy of Sciences, Moscow.

Medvedevskikh, Yu. G.
L. M. Lytvynenko Institute of Physico-organic Chemistry and Coal Chemistry.

Mikhaylov, I. A.
Institute of Biochemical Physics, Russian Academy of Sciences, Moscow, Russia.

Mosyurov, S. E.
Moscow State Pedagogical University, Department of Organic Chemistry, Moscow, Russia.

Nifantiev, E. E.
Moscow State Pedagogical University, Department of Organic Chemistry, Moscow, Russia.

Olkhov, A. A.
N. N. Semenov Institute of Chemical Physics, Russian Academy of Sciences, 4 Kosygin str., Moscow, 119991, Russia.

Petrov, A. L.
Emanuel Institute of Biochemical Physics, Russian Academy of Sciences, ul. Kosygina 4, Moscow, 119334 Russia, E-mail: 28en1937@mail.ru / Semenov Institute of Chemical Physics, Russian Academy of Sciences, ul. Kosygina 4, Moscow, 119334 Russia. Tel.: +7 499 137 7249; Fax: +7 495 651 2191; E-mail: icp@chph.ras.ru.

Popov, A. A.
Institute of Biochemical Physics, Russian Academy of Sciences, Moscow, Russia.

Razumovsky, S. D.
M. Emanuel Institute of Biochemical Physics, Russian Academy of Sciences, Moscow.

Rogovina, S. Z.
N.N. Semenov Institute of chemical Physics, Russian Academy of Sciences, 4 Kosygin str. Moscow 119991, Russia.

Rogovsky, V. S.
Department of Molecular pharmacology and radiobiology, Medico-biological Faculty, N. I. Pirogov Russian National Research Medical University (RNRMU), Moscow, Russia.

Rosenfeld, M. A.
N. M. Emanuel Institute of Biochemical Physics, Russian Academy of Sciences, Moscow.

Rusanova, S. N.
Kazan National Research Technological University.

Shimanovsky, N. L.
M. Emanuel Institute of Biochemical Physics, Russian Academy of Sciences, Moscow.

Sofina, S. Yu.
Kazan National Research Technological University.

Staroverova, O. V.
N. N. Semenov Institute of Chemical Physics, RAS.

Stoyanov, O. V.
Kazan National Research Technological University/N. M. Emanuel Institute of Biochemical Physics.

Temnikova, N. E.
Kazan National Research Technological University.

Volodkin, A.A.
Establishment of the Russian Academy of Sciences Institute of biochemical physics of N. M. Emanuelja, 119991 Moscow, Kosygina, 4; Fax: 095 1374101.

Yevchuk, I. Yu.
NAS of Ukraine, Naukova Str., 3a, Lviv, 79053, e-mail: hop_vfh@ukr.net.

Zaikov, G. E.
N. M. Emanuel Institute of Biochemical Physics RAS, Kosygin Str., 4, Moscow, 119334, e-mail: chembio@sky.chph.ras.ru; E-mail: GEZaikov@Yahoo.com.

Zenitova, L. A.
Kazan National Research Technological University/N. M. Emanuel Institute of Biochemical Physics.

LIST OF ABBREVIATIONS

BBR	Bromated butyl rubber
CBR	Chlorine-butyl rubber
CEP	Chlorinated ethylene-propylene
CEPDC	Chlorine-containing ethylene-propylene-diene cauotchoucs
CFD	Computational fluid dynamics
CSP	Chlorosulfonated polyethylene
CP	Chlorinated polyethylene
DHQ	Dihydroquercetin
DSC	Differential scanning calorimetry
EHF	Extremely high frequency
EPDC	Ethylene-propylene-diene cauotchoucs
EVA	Ethylene with vinylacetate
HF	High frequency
IMD	Implanted medical device
ISM	Industrial, scientific and medical
LF	Low frequency
MEMS and NEMS	Micro- and nanoscale electromechanical systems
MF	Medium frequency
MSC	Mesenchymal stem cells
NMR	Nuclear magnetic resonance
PHAs	Poly(ß-Hydroxyalkanoates)
PHB	Polyhydroxybutyrate
PHBV	Poly(3-Hydroxybutyrate-co-3-Hydroxyvalerate)
PLGAs	Polylactide-co-glycolides
POSS	PCU (Polyhedral oligomeric silsesquioxane and copolymer carbonate-urea urethane
PSD	Pore-size distribution
SAGD	Steam-assisted gravity drainage
SHF	Super high frequency
UC	Urothelial cells
UHF	Ultra high frequency
VLF	Very low frequency
VHF	Very high frequency
WVR	Water vapor resistance
WHO	World health organization

LIST OF SYMBOLS

A	Heat transfer surface area
h_2	Coefficient of heat transfer
Nu	Nusselt number
Pr	Prandtl number
$r(m)$	Radius
r_{max}	Overall largest capillary
Re	Reynolds number
S	Pore saturation
T_0	A reference temperature

GREEK SYMBOLS

$\overline{v}_g$	Speed of the gaseous phase
$v(ms^{-1})$	Fluid velocity
y_F	Relative humidity of fiber of fabric
y_A	Relative humidity of air in pores of fabric
α	Solubility of ozone in water
$\gamma(r)$	Pore volume density function.
μ_g	Viscosity of the gaseous phase
ρ_0	Bulk density of dry material
σ	Surface tension
τ	Tortuosity factor of capillary paths
ψ	Relative humidity

PREFACE

The aim of this book is to provide both a rigorous view and a more practical, understandable view of analytical tools and industrial applications for chemical and polymer engineering graduate students and scientists in related fields.

This volume is structured into different parts devoted to chemical and industrial process, advanced polymers, nanotechnology and drug delivery systems and their applications. Each chapter of the book has been expanded, where relevant, to take account of significant new discoveries and realizations of the importance of key concepts. Furthermore, emphases are placed on the underlying fundamentals and on acquisition of a broad and comprehensive grasp of the field as a whole.

This book brings together research contributions from eminent experts on subjects that have gained prominence in material and chemical engineering and science.

The uniqueness of the topics presented in the book can be gauged from their fundamental and practical importance, particularly latest developments in advanced materials chemical domains.

With contributions from experts from both industry and academia, this book presents the latest developments in the identified areas. This book incorporates appropriate case studies, explanatory notes, and schematics for more clarity and better understanding.

— **Slavcho Kirillov Rakovsky, DSc, Ryszard Kozlowski, PhD and Nekane Guarrotxena, PhD**

CHAPTER 1

METHODOLOGIES ON QUANTUM-CHEMICAL CALCULATION

V. A. BABKIN and G. E. ZAIKOV

CONTENTS

1.1 PART 1—QUANTUM-CHEMICAL CALCULATION OF MOLECULE 1,4-DIMETHYLENECYCLOHEXANE BY METHOD MNDO

1.1.1 INTRODUCTION

For the first time quantum chemical calculation of a molecule of 1,4-di-methylencyclohexane is executed by method MNDO with optimization of geometry on all parameters. The optimized geometrical and electronic structure of this compound is received. Acid power of 1,4-dimethylency-clohexane is theoretically appreciated. It is established, than it to relate to a class of very weak H-acids (pKa = +36, where pKa-universal index of acidity).

The aim of this work is a study of electronic structure of molecule 1,4-dimethylenecyclohexane [1] and theoretical estimation its acid power by quantum-chemical method MNDO. The calculation was done with optimization of all parameters by standard gradient method built-in in PC GAMESS [2]. The calculation was executed in approach the insulated molecule in gas phase. Program MacMolPlt was used for visual presentation of the model of the molecule. [3].

1.1.2 METHODOLOGY

Geometric and electronic structures, general and electronic energies of molecule 1,4-dimethylenecyclohexane was received by method MNDO and are shown on Fig. 1 and in Table 1. The universal factor of acidity was calculated by formula: $pKa = 42.11 - 147.18 * q_{max}^{H+}$ [4, 5] (where, q_{max}^{H+} – a maximum positive charge on atom of the hydrogen $q_{max}^{H+} = +0.04$ (for 1,4-dimethylenecyclohexane q_{max}^{H+} alike Table. 1)). This same formula is used in references [6–16]. $pKa = 36$.

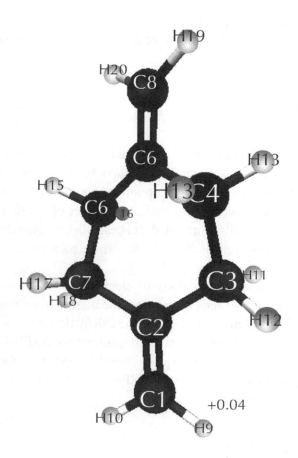

FIGURE 1   Geometric and electronic molecule structure of 1,4-dimethylenecyclohexane.

$$(E_0 = -114880 \text{ kDg/mol}, E_{el} = -559931 \text{ kDg/mol})$$

TABLE 1   Optimized bond lengths, valence corners and charges on atoms of the molecule 1,4-dimethylenecyclohexane.

Bond lengths	R,A	Valence corners	Grad	Atom	Charges on atoms
C(2)-C(1)	1.35	C(1)-C(2)-C(3)	121	C(1)	−0.03
C(3)	1.51	C(2)-C(3)-C(4)	115	C(2)	−0.14

TABLE 1 *(Continued)*

C(4)	1.54	C(3)-C(4)-C(5)	115	C(3)	+0.04
C(5)	1.51	C(4)-C(5)-C(6)	117	C(4)	+0.04
C(6)	1.51	C(2)-C(7)-C(6)	115	C(5)	-0.15
C(6)	1.54	C(1)-C(2)-C(7)	122	C(6)	+0.04
C(7)	1.51	C(4)-C(5)-C(8)	122	C(7)	+0.04
C(8)	1.35	C(2)-C(1)-H(9)	124	C(8)	−0.04
H(9)	1.09	C(2)-C(1)-H(10)	124	H(9)	+0.04
H(10)	1.09	C(2)-C(3)-H(11)	109	H(10)	+0.04
H(11)	1.12	C(2)-C(3)-H(12)	110	H(11)	+0.01
H(12)	1.11	C(3)-C(4)-H(13)	108	H(12)	0.00
H(13)	1.11	C(3)-C(4)-H(14)	109	H(13)	0.00
H(14)	1.12	C(5)-C(6)-H(15)	110	H(14)	+0.01
H(15)	1.11	C(5)-C(6)-H(16)	109	H(15)	0.00
H(16)	1.12	C(2)-C(7)-H(17)	108	H(16)	+0.01
H(17)	1.12	C(2)-C(7)-H(18)	110	H(17)	+0.01
H(18)	1.11	C(5)-C(8)-H(19)	124	H(18)	0.00
H(19)	1.09	C(5)-C(8)-H(20)	124	H(19)	+0.04
H(20)	1.09			H(20)	+0.04

Quantum-chemical calculation of molecule 1,4-dimethylenecyclohexane by method MNDO was executed for the first time. Optimized geometric and electronic structure of this compound was received. Acid power of molecule 1,4-dimethylenecyclohexane was theoretically evaluated

(pKa = 36). This compound pertains to class of very weak H-acids ($pKa > 14$).

1.2 PART 2—QUANTUM-CHEMICAL CALCULATION OF MOLECULE 1-METHYLENE-4-VINYLCYCLOHEXANE BY METHOD MNDO

1.2.1 INTRODUCTION

For the first time quantum chemical calculation of a molecule of 1-methylene-4-vinylcyclohexane is executed by method MNDO with optimization of geometry on all parameters. The optimized geometrical and electronic structure of this compound is received. Acid power of 1-methylen-4-vinylcoclohexane is theoretically appreciated. It is established, than it to relate to a class of very weak H-acids (pKa = +35, where pKa-universal index of acidity).

The aim of this work is a study of electronic structure of molecule 1-methylene-4-vinylcyclohexane [1] and theoretical estimation its acid power by quantum-chemical method MNDO. The calculation was done with optimization of all parameters by standard gradient method built-in in PC GAMESS [2]. The calculation was executed in approach the insulated molecule in gas phase. Program MacMolPlt was used for visual presentation of the model of the molecule [3].

1.2.2 METHODOLOGY

Geometric and electronic structures, general and electronic energies of molecule 1-methylene-4-vinylcyclohexane was received by method MNDO and are shown on Fig. 2 and in Table.2. The universal factor of acidity was calculated by formula: $pKa = 42.11 - 147.18 * q_{max}^{H^+}$ [4, 5] (where, $q_{max}^{H^+}$ – a maximum positive charge on atom of the hydrogen $q_{max}^{H^+} = +0.05$ (for 1-methylene-4-vinylcyclohexane $q_{max}^{H^+}$ alike Table 2)). This same formula is used in references [6–16]. $pKa = 35$.

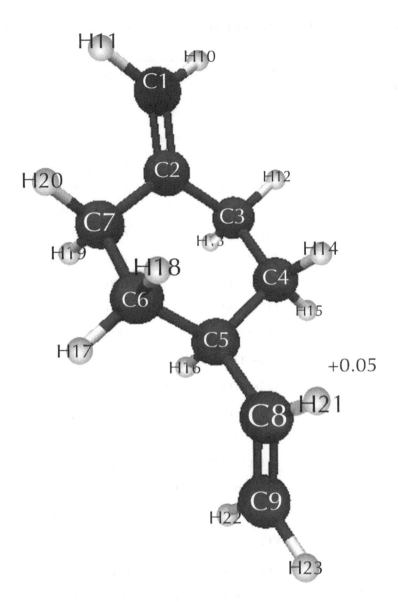

FIGURE 2 Geometric and electronic molecule structure of 1-methylene-4-vinylcyclohexane.

$(E_0 = -129929 \text{ kDg/mol}, E_{el} = -676811 \text{ kDg/mol})$

TABLE 2   Optimized bond lengths, valence corners and charges on atoms of the molecule 1-methylene-4-vinylcyclohexane.

Bond lengths	R,A	Valence corners	Grad	Atom	Charges on atoms
C(2)-C(1)	1.35	C(1)-C(2)-C(3)	122	C(1)	−0.03
C(3)-C(2)	1.51	C(2)-C(3)-C(4)	114	C(2)	−0.15
C(4)-C(3)	1.54	C(3)-C(4)-C(5)	114	C(3)	+0.04
C(5)-C(4)	1.55	C(4)-C(5)-C(6)	112	C(4)	0.00
C(6)-C(5)	1.55	C(2)-C(7)-C(6)	114	C(5)	0.00
C(6)-C(7)	1.54	C(1)-C(2)-C(7)	122	C(6)	0.00
C(7)-C(2)	1.51	C(4)-C(5)-C(8)	111	C(7)	+0.04
C(8)-C(5)	1.51	C(5)-C(8)-C(9)	126	C(8)	−0.11
C(9)-C(8)	1.34	C(2)-C(1)-H(10)	124	C(9)	−0.06
H(10)-C(1)	1.09	C(2)-C(1)-H(11)	124	H(10)	+0.04
H(11)-C(1)	1.09	C(2)-C(3)-H(12)	111	H(11)	+0.04
H(12)-C(3)	1.11	C(2)-C(3)-H(13)	109	H(12)	0.00
H(13)-C(3)	1.12	C(3)-C(4)-H(14)	109	H(13)	+0.01
H(14)-C(4)	1.11	C(3)-C(4)-H(15)	109	H(14)	+0.01
H(15)-C(4)	1.11	C(4)-C(5)-H(16)	107	H(15)	+0.01
H(16)-C(5)	1.12	C(5)-C(6)-H(17)	109	H(16)	+0.01
H(17)-C(6)	1.11	C(5)-C(6)-H(18)	110	H(17)	+0.01
H(18)-C(6)	1.11	C(2)-C(7)-H(19)	109	H(18)	+0.01
H(19)-C(7)	1.12	C(2)-C(7)-H(20)	111	H(19)	+0.01
H(20)-C(7)	1.11	C(5)-C(8)-H(21)	115	H(20)	0.00
H(21)-C(8)	1.10	C(8)-C(9)-H(22)	125	**H(21)**	**+0.05**
H(22)-C(9)	1.09	C(8)-C(9)-H(23)	122	H(22)	+0.04
H(23)-C(9)	1.09			H(23)	+0.04

Quantum-chemical calculation of molecule 1-methylene-4-vinylcy-clohexane by method MNDO was executed for the first time. Optimized geometric and electronic structure of this compound was received. Acid power of molecule 1-methylene-4-vinylcyclohexane was theoretically evaluated (pKa = 35). This compound pertains to class of very weak H-acids (pKa > 14).

1.3 PART 3—QUANTUM-CHEMICAL CALCULATION OF MOLECULE METHYLENCYCLOOCTANE BY METHOD MNDO

1.3.1 INTRODUCTION

For the first time quantum chemical calculation of a molecule of methylencyclooctane is executed by method MNDO with optimization of geometry on all parameters. The optimized geometrical and electronic structure of this compound is received. Acid power of methylencyclooctane is theoretically appreciated. It is established, than it to relate to a class of very weak H-acids (pKa=+36, where pKa-universal index of acidity).

The aim of this work is a study of electronic structure of molecule methylencyclooctane [1] and theoretical estimation its acid power by quantum-chemical method MNDO. The calculation was done with optimization of all parameters by standard gradient method built-in in PC GAMESS [2]. The calculation was executed in approach the insulated molecule in gas phase. Program MacMolPlt was used for visual presentation of the model of the molecule. [3].

1.3.1 METHODOLOGY

Geometric and electronic structures, general and electronic energies of molecule methylencyclooctane was received by method MNDO and are shown on Fig. 3 and in Table.3. The universal factor of acidity was calculated by formula: pKa = 42.11–147.18*q_{max}^{H+} [4, 5] (where, q_{max}^{H+} – a maximum positive charge on atom of the hydrogen q_{max}^{H+}= +0.04 (for methylencyclooctane $q_{max}^{H^+}$ alike Table 3)). This same formula is used in references [6–16]. pKa = 36.

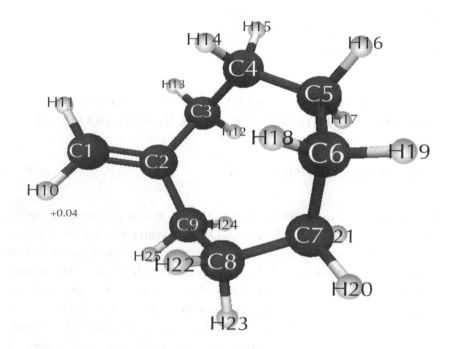

FIGURE 3 Geometric and electronic molecule structure of methylencyclooctane.

$$(E_0 = -132969 \text{ kDg/mol}, E_{el} = -734560 \text{ kDg/mol})$$

TABLE 3 Optimized bond lengths, valence corners and charges on atoms of the molecule methylencyclooctane.

Bond lengths	R,A	Valence corners	Grad	Atom	Charges on atoms
C(1)-C(2)	1.35	C(2)-C(1)-H(10)	124	C(1)	−0.04
C(2)-C(3)	1.52	C(2)-C(1)-H(11)	124	C(2)	−0.15
C(3)-C(4)	1.54	C(2)-C(3)-H(12)	110	C(3)	0.04
C(4)-C(5)	1.54	C(2)-C(3)-H(13)	109	C(4)	−0.01
C(5)-C(6)	1.54	C(2)-C(9)-H(24)	110	C(5)	−0.01

TABLE 3 *(Continued)*

C(6)-C(7)	1.54	C(2)-C(9)-H(25)	109	C(6)	−0.00
C(7)-C(8)	1.54	C(3)-C(4)-H(14)	110	C(7)	−0.01
C(8)-C(9)	1.54	C(3)-C(4)-H(15)	107	C(8)	−0.01
C(9)-C(2)	1.52	C(4)-C(3)-H(12)	110	C(9)	0.04
H(10)-C(1)	1.09	C(4)-C(3)-H(13)	107	H(10)	0.04
H(11)-C(1)	1.09	C(4)-C(5)-H(16)	107	H(11)	0.04
H(12)-C(3)	1.11	C(4)-C(5)-H(17)	110	H(12)	0.01
H(13)-C(3)	1.12	C(5)-C(4)-H(14)	110	H(13)	0.00
H(14)-C(4)	1.11	C(5)-C(4)-H(15)	107	H(14)	0.01
H(15)-C(4)	1.12	C(5)-C(6)-H(18)	110	H(15)	−0.00
H(16)-C(5)	1.12	C(5)-C(6)-H(19)	107	H(16)	−0.00
H(17)-C(5)	1.12	C(6)-C(5)-H(16)	107	H(17)	0.01
H(18)-C(6)	1.11	C(6)-C(5)-H(17)	110	H(18)	0.01
H(19)-C(6)	1.12	C(6)-C(7)-H(20)	107	H(19)	−0.00
H(20)-C(7)	1.12	C(6)-C(7)-H(21)	110	H(20)	−0.00
H(21)-C(7)	1.11	C(7)-C(6)-H(18)	110	H(21)	0.01
H(22)-C(8)	1.11	C(7)-C(6)-H(19)	107	H(22)	0.01
H(23)-C(8)	1.12	C(7)-C(8)-H(22)	110	H(23)	−0.00
H(24)-C(9)	1.11	C(7)-C(8)-H(23)	107	H(24)	0.01
H(25)-C(9)	1.12	C(8)-C(7)-H(20)	107	H(25)	0.00
		C(8)-C(7)-H(21)	110		
		C(8)-C(9)-H(24)	110		
		C(8)-C(9)-H(25)	107		
		C(9)-C(8)-H(22)	110		
		C(9)-C(8)-H(23)	107		

1.4 PART 4—QUANTUM-CHEMICAL CALCULATION OF MOLECULE BICYCLO[3,1,0]HEXANE BY METHOD *AB INITIO*

1.4.1 INTRODUCTION

For the first time quantum chemical calculation of a molecule of bicyclo[3,1,0]hexane is executed by method ab initio in base 6–311G** with optimization of geometry on all parameters for the first time. The optimized geometrical and electronic structure of this compound is received. Acid power of bicyclo [3,1,0] hexane is theoretically appreciated. It is established, than it to relate to a class of very weak H-acids (pKa = +34, where pKa-universal index of acidity).

The aim of this work is a study of electronic structure of molecule bicyclo[3,1,0]hexane [1] and theoretical estimation its acid power by quantum-chemical method *ab initio* in base 6–311G**. The calculation was done with optimization of all parameters by standard gradient method built-in in PC GAMESS [2]. The calculation was executed in approach the insulated molecule in gas phase. Program MacMolPlt was used for visual presentation of the model of the molecule. [3].

1.4.2 METHODOLOGY

Geometric and electronic structures, general and electronic energies of molecule bicyclo[3,1,0]hexane was received by method *ab initio* in base 6–311G** and are shown on Fig. 4 and in Table 4. The universal factor of acidity was calculated by formula: $pKa = 49.04 - 134.6*q_{max}^{H^+}$ [4,5] (where, $q_{max}^{H^+}$ – a maximum positive charge on atom of the hydrogen $q_{max}^{H^+} = +0.11$ (for bicyclo[3,1,0]hexane $q_{max}^{H^+}$ alike Table 4)). This same formula is used in references [6–16]. $pKa = 34$.

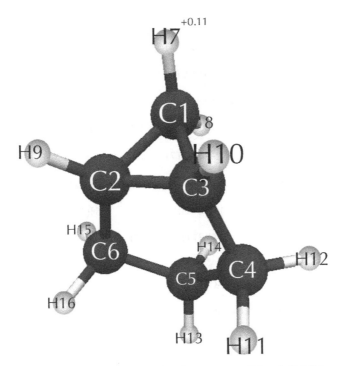

FIGURE 4 Geometric and electronic molecule structure of bicyclo[3,1,0]hexane.

$(E_0 = -610802 \text{ kDg/mol}, E_{el} = -1253253 \text{ kDg/mol})$

TABLE 4 Optimized bond lengths, valence corners and charges on atoms of the molecule bicyclo [3,1,0] hexane.

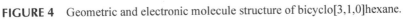

Bond lengths	R,A	Valence corners	Grad	Atom	Charges on atoms
C(2)-C(1)	1.50	C(3)-C(2)-C(1)	60	C(1)	−0.18
C(2)-C(3)	1.50	C(1)-C(3)-C(2)	60	C(2)	−0.19
C(3)-C(1)	1.50	C(4)-C(3)-C(2)	108	C(3)	−0.19
C(4)-C(3)	1.52	C(2)-C(1)-C(3)	60	C(4)	−0.09
C(4)-C(5)	1.54	C(5)-C(4)-C(3)	105	C(5)	−0.27
C(5)-C(6)	1.54	C(1)-C(3)-C(4)	118	C(6)	−0.09

TABLE 4   *(Continued)*

C(6)-C(2)	1.52	C(6)-C(5)-C(4)	105	H(7)	+0.11
H(7)-C(1)	1.08	C(2)-C(6)-C(5)	105	H(8)	+0.11
H(8)-C(1)	1.08	C(1)-C(2)-C(6)	118	H(9)	+0.11
H(9)-C(2)	1.08	C(3)-C(2)-C(6)	108	H(10)	+0.11
H(10)-C(3)	1.08	C(2)-C(1)-H(7)	117	H(11)	+0.09
H(11)-C(4)	1.09	C(2)-C(1)-H(8)	120	H(12)	+0.10
H(12)-C(4)	1.09	C(1)-C(2)-H(9)	118	H(13)	+0.10
H(13)-C(5)	1.08	C(3)-C(2)-H(9)	121	H(14)	+0.10
					+0.10
H(14)-C(5)	1.09	C(1)-C(3)-H(10)	118	H(15)	+0.09
H(15)-C(6)	1.09	C(3)-C(4)-H(11)	109	H(16)	
H(16)-C(6)	1.09	C(5)-C(4)-H(11)	110		
		C(3)-C(4)-H(12)	113		
		C(5)-C(4)-H(12)	112		
		C(6)-C(5)-H(13)	112		
		C(6)-C(5)-H(14)	110		
		C(2)-C(6)-H(15)	113		
		C(2)-C(6)-H(16)	109		

Quantum-chemical calculation of molecule bicyclo[3,1,0]hexane by method *ab initio* in base 6–311G** was executed for the first time. Optimized geometric and electronic structure of this compound was received. Acid power of molecule bicyclo[3,1,0]hexane was theoretically evaluated (*pKa* = 34). This compound pertains to class of very weak H-acids (*pKa* > 14).

1.5 PART 5—QUANTUM-CHEMICAL CALCULATION OF MOLECULE BICYCLO [4,1,0] HEPTANE BY METHOD *AB INITIO*

1.5.1 INTRODUCTION

For the first time quantum chemical calculation of a molecule of bicyclo [4,1,0] heptane is executed by method *ab initio* in base 6–311G** with optimization of geometry on all parameters for the first time. The optimized geometrical and electronic structure of this compound is received. Acid power of bicyclo [4,1,0] heptane is theoretically appreciated. It is established, than it to relate to a class of very weak H-acids (*pKa*=+34, where pKa-universal index of acidity).

The aim of this work is a study of electronic structure of molecule bicyclo[4,1,0]heptane [1] and theoretical estimation its acid power by quantum-chemical method *ab initio* in base 6–311G**. The calculation was done with optimization of all parameters by standard gradient method built-in in PC GAMESS [2]. The calculation was executed in approach the insulated molecule in gas phase. Program MacMolPlt was used for visual presentation of the model of the molecule. [3].

1.5.2 METHODOLOGY

Geometric and electronic structures, general and electronic energies of molecule bicyclo[4,1,0]heptane was received by method ab initio in base 6–311G** and are shown on Fig. 5 and in Table 5. The universal factor of acidity was calculated by formula: pKa = 49.04–134.6*q_{max}^{H+} [4, 5] (where, q_{max}^{H+} – a maximum positive charge on atom of the hydrogen q_{max}^{H+} = +0.11 (for bicyclo [4,1,0] heptane q_{max}^{H+} alike Table 5)). This same formula is used in references [6—16]. pKa = 34.

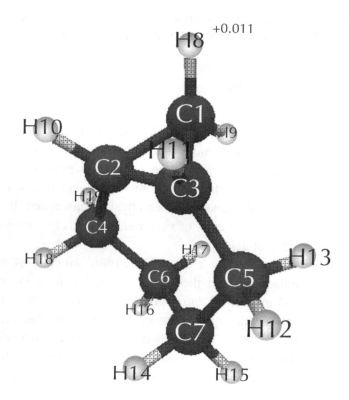

FIGURE 5   Geometric and electronic molecule structure of bicyclo[4,1,0]heptane.

$(E_0 = -713133 \text{ kDg/mol}, E_{el} = -1546260 \text{ kDg/mol})$

TABLE 5   Optimized bond lengths, valence corners and charges on atoms of the molecule bicyclo[4,1,0]heptane.

Bond lengths	R,A	Valence corners	Grad	Atom	Charges on atoms
C(2)-C(1)	1.50	C(3)-C(2)-C(1)	60	C(1)	−0.18
C(2)-C(3)	1.51	C(1)-C(3)-C(2)	60	C(2)	−0.17
C(3)-C(1)	1.50	C(5)-C(3)-C(2)	120	C(3)	−0.19
C(4)-C(2)	1.52	C(2)-C(1)-C(3)	60	C(4)	−0.11
C(5)-C(3)	1.53	C(7)-C(5)-C(3)	113	C(5)	−0.15

TABLE 5   *(Continued)*

C(5)-C(7)	1.53	C(1)-C(2)-C(4)	122	C(6)	−0.23
C(6)-C(4)	1.53	C(3)-C(2)-C(4)	120	C(7)	−0.16
C(7)-C(6)	1.53	C(1)-C(3)-C(5)	120	H(8)	+0.11
H(8)-C(1)	1.08	C(6)-C(7)-C(5)	112	H(9)	+0.11
H(9)-C(1)	1.08	C(2)-C(4)-C(6)	113	H(10)	+0.10
H(10)-C(2)	1.08	C(4)-C(6)-C(7)	111	H(11)	+0.11
H(11)-C(3)	1.08	C(2)-C(1)-H(8)	118	H(12)	+0.10
H(12)-C(5)	1.09	C(2)-C(1)-H(9)	118	H(13)	+0.10
H(13)-C(5)	1.09	C(1)-C(2)-H(10)	115	H(14)	+0.09
H(14)-C(7)	1.09	C(3)-C(2)-H(10)	116	H(15)	+0.09
H(15)-C(7)	1.09	C(1)-C(3)-H(11)	115	H(16)	+0.10
H(16)-C(6)	1.09	C(3)-C(5)-H(12)	109	H(17)	+0.09
H(17)-C(6)	1.09	C(7)-C(5)-H(12)	109	H(18)	+0.09
H(18)-C(4)	1.09	C(3)-C(5)-H(13)	110	H(19)	+0.10
H(19)-C(4)	1.09	C(7)-C(5)-H(13)	109		
		C(6)-C(7)-H(14)	109		
		C(6)-C(7)-H(15)	111		
		C(4)-C(6)-H(16)	109		
		C(4)-C(6)-H(17)	111		
		C(2)-C(4)-H(18)	108		
		C(2)-C(4)-H(19)	110		

Quantum-chemical calculation of molecule bicyclo[4,1,0]heptane by method *ab initio* in base 6–311G** was executed for the first time. Optimized geometric and electronic structure of this compound was received. Acid power of molecule bicyclo[4,1,0]heptane was theoretically evaluated (pKa = 34.). This compound pertains to class of very weak H-acids (pKa>14).

1.6 PART 6—QUANTUM-CHEMICAL CALCULATION OF MOLECULE 1,3-DIPHENYLINDENE BY METHOD *AB INITIO*

1.6.1 INTRODUCTION

For the first time quantum chemical calculation of a molecule of 1,3-diphenylindene is executed by method *ab initio* with optimization of geometry on all parameters. The optimized geometrical and electronic structure of this compound is received. Acid power of 1,3-diphenylindene is theoretically appreciated. It is established, than it to relate to a class of very weak H-acids (pKa = +30, where pKa-universal index of acidity).

     The aim of this work is a study of electronic structure of molecule 1,3-diphenylindene [1] and theoretical estimation its acid power by quantum-chemical method *ab initio* in base 6–311G**. The calculation was done with optimization of all parameters by standard gradient method built-in in PC GAMESS [2]. The calculation was executed in approach the insulated molecule in gas phase. Program MacMolPlt was used for visual presentation of the model of the molecule. [3].

1.6.2 METHODOLOGY

Geometric and electronic structures, general and electronic energies of molecule 1,3-diphenylindene was received by method ab initio in base 6–311G** and are shown on Fig. 6. and in Table 6. The universal factor of acidity was calculated by formula: pKa = 49.04 – 134.6*q_{max}^{H+} [4, 5] (where, q_{max}^{H+} – a maximum positive charge on atom of the hydrogen q_{max}^{H+}= +0.14 (for 1,3-diphenylindene q_{max}^{H+} alike Table 6)). This same formula is used in references [6–16]. pKa = 30.

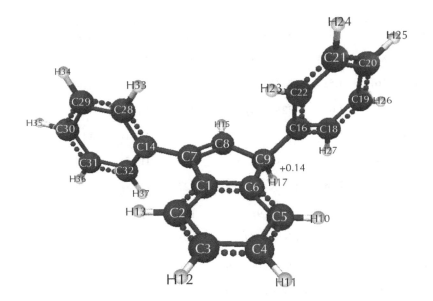

FIGURE 6 Geometric and electronic molecule structure of 1,3-diphenylindene.

$$(E_0 = -2112398 \text{ kDg/mol}, E_{el} = -5939871 \text{ kDg/mol})$$

TABLE 6 Optimized bond lengths, valence corners and charges on atoms of the molecule 1,3-diphenylindene.

Bond lengths	R,A	Valence corners	Grad	Atom	Charges on atoms
C(2)-C(1)	1.38	C(5)-C(6)-C(1)	121	C(1)	0.00
C(3)-C(2)	1.39	C(9)-C(6)-C(1)	109	C(2)	−0.07
C(4)-C(3)	1.39	C(14)-C(7)-C(1)	124	C(3)	−0.09
C(5)-C(4)	1.39	C(1)-C(2)-C(3)	118	C(4)	−0.09
C(6)-C(5)	1.38	C(2)-C(3)-C(4)	121	C(5)	−0.05
C(6)-C(1)	1.39	C(3)-C(4)-C(5)	121	C(6)	−0.11
C(6)-C(9)	1.52	C(9)-C(6)-C(5)	130	C(7)	0.00
C(7)-C(1)	1.48	C(4)-C(5)-C(6)	119	C(8)	−0.07
C(7)-C(14)	1.49	C(2)-C(1)-C(6)	121	C(9)	−0.06

TABLE 6 *(Continued)*

C(8)-C(7)	1.33	C(8)-C(9)-C(6)	101	H(10)	+0.09
C(9)-C(8)	1.52	C(16)-C(9)-C(6)	116	H(11)	+0.09
C(9)-C(16)	1.52	C(2)-C(1)-C(7)	131	H(12)	+0.09
H(10)-C(5)	1.08	C(32)-C(14)-C(7)	120	H(13)	+0.10
H(11)-C(4)	1.08	C(28)-C(14)-C(7)	121	C(14)	−0.11
H(12)-C(3)	1.08	C(1)-C(7)-C(8)	109	H(15)	+0.09
H(13)-C(2)	1.08	C(14)-C(7)-C(8)	127	C(16)	−0.10
C(14)-C(32)	1.39	C(16)-C(9)-C(8)	115	H(17)	+0.14
H(15)-C(8)	1.07	C(7)-C(8)-C(9)	112	C(18)	−0.08
C(16)-C(22)	1.39	C(22)-C(16)-C(9)	121	C(19)	−0.09
H(17)-C(9)	1.09	C(18)-C(16)-C(9)	120	C(20)	−0.10
C(18)-C(16)	1.39	C(4)-C(5)-H(10)	120	C(21)	−0.09
C(19)-C(18)	1.39	C(3)-C(4)-H(11)	120	C(22)	−0.05
C(20)-C(19)	1.38	C(2)-C(3)-H(12)	120	H(23)	+0.10
C(21)-C(20)	1.39	C(1)-C(2)-H(13)	121	H(24)	+0.10
C(22)-C(21)	1.38	C(31)-C(32)-C(14)	121	H(25)	+0.10
H(23)-C(22)	1.08	C(7)-C(8)-H(15)	126	H(26)	+0.09
H(24)-C(21)	1.08	C(21)-C(22)-C(16)	121	H(27)	+0.09
H(25)-C(20)	1.08	C(8)-C(9)-H(17)	108	C(28)	−0.06
H(26)-C(19)	1.08	C(16)-C(9)-H(17)	108	C(29)	−0.09
H(27)-C(18)	1.08	C(22)-C(16)-C(18)	118	C(30)	−0.10
C(28)-C(14)	1.39	C(16)-C(18)-C(19)	121	C(31)	−0.09
C(29)-C(28)	1.39	C(18)-C(19)-C(20)	120	C(32)	−0.06
C(30)-C(29)	1.38	C(19)-C(20)-C(21)	119	H(33)	+0.09
C(31)-C(30)	1.38	C(20)-C(21)-C(22)	120	H(34)	+0.10
C(32)-C(31)	1.38	C(21)-C(22)-H(23)	120	H(35)	+0.10

TABLE 6 *(Continued)*

H(33)-C(28)	1.08	C(20)-C(21)-H(24)	120	H(36)	+0.10
H(34)-C(29)	1.08	C(19)-C(20)-H(25)	120	H(37)	+0.09
H(35)-C(30)	1.08	C(18)-C(19)-H(26)	120		
H(36)-C(31)	1.08	C(16)-C(18)-H(27)	120		
H(37)-C(32)	1.08	C(32)-C(14)-C(28)	119		
		C(14)-C(28)-C(29)	121		
		C(28)-C(29)-C(30)	120		
		C(29)-C(30)-C(31)	120		
		C(30)-C(31)-C(32)	120		
		C(14)-C(28)-H(33)	120		
		C(28)-C(29)-H(34)	120		
		C(29)-C(30)-H(35)	120		
		C(30)-C(31)-H(36)	120		
		C(31)-C(32)-H(37)	120		

Quantum-chemical calculation of molecule 1,3-diphenylindene by method *ab initio* in base 6 − 311G** was executed for the first time. Optimized geometric and electronic structure of this compound was received. Acid power of molecule 1,3-diphenylindene was theoretically evaluated (pKa = 30.). This compound pertains to class of very weak H-acids (pKa > 14).

1.7 PART 7—QUANTUM-CHEMICAL CALCULATION OF MOLECULE 3,3'-DIINDELYL BY METHOD *AB INITIO*

1.7.1 INTRODUCTION

For the first time quantum chemical calculation of a molecule of 3,3'-diindenyl is executed by method *ab initio* with optimization of geometry on all parameters. The optimized geometrical and electronic structure of this compound is received. Acid power of 3,3'-diindenyl is theoretically appreciated. It is established, than it to relate to a class of very weak H-acids (pKa = +32, where pKa-universal index of acidity).

The aim of this work is a study of electronic structure of molecule 3,3'-diindenyl [1] and theoretical estimation its acid power by quantum-chemical method *ab initio* in base 6–311G**. The calculation was done with optimization of all parameters by standard gradient method built-in in PC GAMESS [2]. The calculation was executed in approach the insulated molecule in gas phase. Program MacMolPlt was used for visual presentation of the model of the molecule. [3].

1.7.2   METHODOLOGY

Geometric and electronic structures, general and electronic energies of molecule 3,3'-diindenyl was received by method *ab initio* in base 6–311G** and are shown on Fig. 7 and in Table 7. The universal factor of acidity was calculated by formula: pKa = $49.04 - 134.6 * q_{max}^{H^+}$ [4, 5] (where, $q_{max}^{H^+}$ – a maximum positive charge on atom of the hydrogen $q_{max}^{H^+} = +0.13$ (for 3,3'-diindenyl $q_{max}^{H^+}$ alike Table 7)). This same formula is used in references [6—16]. pKa=32.

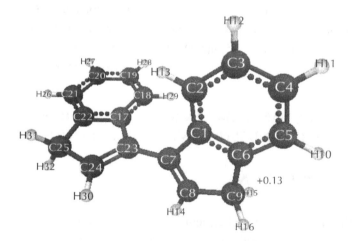

FIGURE 7   Geometric and electronic molecule structure of 3,3'-diindenyl.

$(E_0 = -18\ 11035\ kDg/mol,\ E_{el} = -4873383\ kDg/mol)$

TABLE 7 Optimized bond lengths, valence corners and charges on atoms of the molecule 3,3'-diindenyl.

Bond lengths	R,A	Valence corners	Grad	Atom	Charges on atoms
C(2)-C(1)	1.38	C(5)-C(6)-C(1)	121	C(1)	−0.01
C(3)-C(2)	1.39	C(9)-C(6)-C(1)	109	C(2)	−0.07
C(4)-C(3)	1.39	C(1)-C(2)-C(3)	119	C(3)	−0.09
C(5)-C(4)	1.39	C(2)-C(3)-C(4)	121	C(4)	−0.09
C(6)-C(5)	1.38	C(3)-C(4)-C(5)	121	C(5)	−0.08
C(6)-C(1)	1.39	C(9)-C(6)-C(5)	130	C(6)	−0.14
C(6)-C(9)	1.51	C(4)-C(5)-C(6)	119	C(7)	−0.03
C(8)-C(7)	1.33	C(8)-C(9)-C(6)	102	C(9)	−0.06
C(9)-C(8)	1.51	C(2)-C(1)-C(7)	131	H(10)	+0.08
H(10)-C(5)	1.08	C(1)-C(7)-C(8)	109	H(11)	+0.09
H(11)-C(4)	1.08	C(23)-C(7)-C(8)	127	H(12)	+0.09
H(12)-C(3)	1.08	C(7)-C(8)-C(9)	112	H(13)	+0.10
H(13)-C(2)	1.08	C(4)-C(5)-H(10)	120	H(14)	+0.09
H(14)-C(8)	1.07	C(3)-C(4)-H(11)	120	H(15)	+0.13
H(15)-C(9)	1.09	C(2)-C(3)-H(12)	120	H(16)	+0.13
H(16)-C(9)	1.09	C(1)-C(2)-H(13)	121	C(17)	−0.01
C(17)-C(22)	1.39	C(7)-C(8)-H(14)	125	C(18)	−0.07
C(17)-C(23)	1.48	C(8)-C(9)-H(15)	112	C(19)	−0.09

TABLE 7   *(Continued)*

C(18)-C(17)	1.38	C(8)-C(9)-H(16)	112	C(20)	−0.09
C(19)-C(18)	1.39	C(21)-C(22)-C(17)	121	C(21)	−0.08
C(20)-C(19)	1.39	C(25)-C(22)-C(17)	109	C(22)	−0.14
C(21)-C(20)	1.39	C(7)-C(23)-C(17)	125	C(23)	−0.04
C(22)-C(21)	1.38	C(24)-C(23)-C(17)	109	C(24)	−0.13
C(22)-C(25)	1.51	C(22)-C(17)-C(18)	121	C(25)	−0.06
C(23)-C(7)	1.48	C(23)-C(17)-C(18)	131	H(26)	+0.08
C(24)-C(23)	1.33	C(17)-C(18)-C(19)	119	H(27)	+0.09
C(25)-C(24)	1.51	C(18)-C(19)-C(20)	121	H(28)	+0.09
H(26)-C(21)	1.08	C(19)-C(20)-C(21)	121	H(29)	+0.10
H(27)-C(20)	1.08	C(25)-C(22)-C(21)	130	H(30)	+0.09
H(28)-C(19)	1.08	C(20)-C(21)-C(22)	119	H(31)	+0.13
H(29)-C(18)	1.07	C(24)-C(25)-C(22)	102	H(32)	+0.12
H(30)-C(24)	1.07	C(23)-C(17)-C(22)	108		
H(31)-C(25)	1.09	C(1)-C(7)-C(23)	125		
H(32)-C(25)	1.09	C(7)-C(23)-C(24)	127		
		C(23)-C(24)-C(25)	112		
		C(20)-C(21)-H(26)	120		
		C(19)-C(20)-H(27)	120		
		C(18)-C(19)-H(28)	120		
		C(17)-C(18)-H(29)	121		

TABLE 7 *(Continued)*

C(23)-C(24)-H(30)	125
C(24)-C(25)-H(31)	112
C(24)-C(25)-H(32)	112

Quantum-chemical calculation of molecule 3,3'-diindenyl by method *ab initio* in base 6–311G** was executed for the first time. Optimized geometric and electronic structure of this compound was received. Acid power of molecule 1,3-diphenylindene was theoretically evaluated (pKa = 32). This compound pertains to class of very weak H-acids (pKa > 14).

KEYWORDS

- *Ab initio* **method**
- **Bicyclo[3,1,0]hexane**
- **Bicyclo[4,1,0]heptane**
- **1,3-Diphenylindene**
- **1,4-Dimethylencyclohexane**
- **3,3'-Diindenyl**
- **MacMolPlt program**
- **Methylencyclooctane**
- **1-Methylene-4-vinylcyclohexane**

REFERENCES

1. Kennedi, J. Cationic polimerization of olefins. *Moscow*, **1978**, 431.
2. Shmidt, M. W.; Baldrosge, K. K.; Elbert, J. A.; Gordon, M. S.; Enseh, J. H.; Koseki, S.; Matsvnaga, N.; Nguyen, K. A.; Su, S. J., et al. *J. Comput. Chem.* **1993**, *14,* 1347–1363.

3. Bode, B. M.; Gordon, M. S. J. Mol. *Graphics Mod.*, **1998,** *16,* 133–138.
4. Babkin, V. A.; Fedunov, R. G.; Minsker, K. S., et al. *O xidation communication,* **2002,** *25(1),* 21–47.
5. Babkin, V. A., et al. Oxidation communication, **1998,** *21(4),* 454–460.
6. Babkin, V. A.; Dmitriev, V. Yu.; Zaikov, G. E. Geometrical and electronic structure of molecule benzilpenicillin by method *ab initio.* In:*Quantum-chemical calculations of molecular system as the basis of nanotechnologies in applied quantum chemistry. I.* New York, Nova Publisher, **2012,** 7–10.
7. Babkin, V. A.; Tsykanov, A. B. Geometrical and electronic structure of molecule cellulose by method *ab initio.* In: *Quantum-chemical calculations of molecular system as the basis of nanotechnologies in applied quantum chemistry. I.* New York, Nova Publisher, **2012,** 31–34.
8. Babkin, V. A.; Dmitriev, V. Yu.; Zaikov, G. E. Geometrical and electronic structure of molecule aniline by method *ab initio.* In: Quantum-chemical calculations of molecular system as the basis of nanotechnologies in applied quantum chemistry. *I.* New York, Nova Publisher, **2012,** 89–91.
9. Babkin, V. A.; Dmitriev, V. Yu.; Zaikov, G. E.. Geometrical and electronic structure of molecule butene-1 by method *ab initio.* In: *Quantum-chemical calculations of molecular system as the basis of nanotechnologies in applied quantum chemistry. I.* New York, Nova Publisher, **2012,** 109–111.
10. Babkin, V. A.; Dmitriev, V. Yu.; Zaikov, G. E.. Geometrical and electronic structure of molecule butene-2 by method *ab initio.* In: *Quantum-chemical calculations of molecular system as the basis of nanotechnologies in applied quantum chemistry. I.* New York, Nova Publisher, **2012,** 113–115.
11. Babkin, V. A.; Galenkin, V. V. Geometrical and electronic structure of molecule 3, 3-dimethylbutene-1 by method *ab initio.* In: *Quantum-chemical calculations of molecular system as the basis of nanotechnologies in applied quantum chemistry. I.* New York, Nova Publisher, **2012,** 129–131.
12. Babkin, V. A.; Andreev, D. S. Geometrical and electronic structure of molecule 4, 4-dimethylpentene-1 by method *ab initio.* In: *Quantum-chemical calculations of molecular system as the basis of nanotechnologies in applied quantum chemistry. I.* New York, Nova Publisher, **2012,** 141–143.
13. Babkin, V. A.; Andreev, D. S. Geometrical and electronic structure of molecule 4-methylhexene-1 by method *ab initio.* In: *Quantum-chemical calculations of molecular system as the basis of nanotechnologies in applied quantum chemistry. I.* New York, Nova Publisher, **2012,** 145–147.
14. Babkin, V. A.; Andreev, D. S. Geometrical and electronic structure of molecule 4-methylpentene-1 by method *ab initio.* In: *Quantum-chemical calculations of molecular system as the basis of nanotechnologies in applied quantum chemistry. I.* New York, Nova Publisher, **2012,** 149–151.
15. Babkin, V. A.; Andreev, D. S. Geometrical and electronic structure of molecule isobutylene by method *ab initio.* In: *Quantum-chemical calculations of molecular system as the basis of nanotechnologies in applied quantum chemistry. I.* New York, Nova Publisher, **2012,** 155–157.

16. Babkin, V. A.; Andreev, D. S. Geometrical and electronic structure of molecule 2-methylbutene-1 by method *ab initio*. In: *Quantum-chemical calculations of molecular system as the basis of nanotechnologies in applied quantum chemistry. I.* New York, Nova Publisher, **2012**, 159–161.

CHAPTER 2

REACTIVITY OF TERT.BUTYL ESTER 3-(3', 5'-DI-TERT.BUTYL-4'-HYDROXYPHENYL)-PROPIONIC ACID IN THE REACTIONS OF OXIDATION

A. A. VOLODKIN, G. E. ZAIKOV, N. M. EVTEEVA, S. M. LOMAKIN, and N. N. MADYUSKIN

CONTENTS

2.1   INTRODUCTION

It is shown that *tert*-butyl ester 3-(3', 5'-*di-tert.*butyl-4'-hydroxyphenyl)-propionic acid possesses anomalously high antioxidative efficacy in radical reaction in the conditions of the initiated oxidation of *iso*-propyl benzene with an inhibition constant $K_7 = 3 \times 10^6$ l·mol⁻¹·s⁻¹ and number of stopping of chain f = 5. On the basis of results of quantum-chemical calculations enthalpies (H_f°), entropies (S_f°) and energy of a cleavage of communications OH-bond of *tert.*butyl and methyl ester 3-(3', 5'-*di-tert.* butyl-4'-hydroxyphenyl)-propionic acid are calculated energy of formation ($-E_f^\circ$).

In the conditions of the initiated oxidation of iso-propyl benzene (ArH) it is formed peroxy radical ArOO·, which participates in continuation of a chain of considerably-chain oxidation [1–3]. Specific reaction rates peroxy radicals with phenolic antioxidants (κ_7) are in limens 10^3–10^5 l·mol⁻¹·s⁻¹, and values of dissociation energy O–H depend on critical increment of energy of this reaction [3–5]. The linear relation of sizes of energy of a cleavage of communications O–H (D_{OH}) bond from constants κ_7 [6] is positioned that has allowed to use settlement procedures for forecasting of efficacy of antioxidants from their structures [7, 8]. However according to Ref. [9] reactions of some derivatives oxynaphtochinone with hydroperoxy radical are reversible also balance it is displaced towards initial components. This data will not be coordinated with the theory inhibiting oxidations on which reaction of radical RO_2 with inhibitor (InOH) is reverse [3]. The analysis of the data on research inhibiting oxidation does not allow to character is unequivocally efficacy of an antioxidant on set of parameters: k_7 and to number of stopping of chain (f). There was a necessity of the account of the structure factors following from geometrical and power parameters of an antioxidant.

In the present work it is obtained *tert.*butyl ester 3-(3', 5'-di-tert-butyl-4'-hydroxyphenyl)-propionic acid. In reaction inhibiting oxidations of *iso*-propylbenzene the specified antioxidant reacts with ArOO · with $k_7 = 3 \times 10^6$ l·mol⁻¹·s⁻¹ and f = 5. From quantum-chemical calculations in approach PM6 geometrical and power parameters of structures methyl and tert-butyl ester 3-(3', 5'-di-tert-butyl-4'-hydroxyphenyl)-propionic acids are calculated and the results considering transformations of mediate structures at interaction of substrate with peroxy radical are made.

2.2 EXPERIMENTAL PART

^{1}H spectrums of the initial and received compounds registered NMR on the device "Bruker WM-400" concerning a signal of residual protons. IR-spectra wrote down on a spectrometer "PERKIN-ELMER 1725-X" in crystals a method of diffusive reflectance.

Calculation of parameters of structures of the received compounds made on a method of Hartrii-Fock (UHF), using program MOPAC 2009 [10]; mathematical processing of results carried out in program Origin 6.1.

Methyl ester 3-(3', 5'-di-tert-butyl-4'-hydroxyphenyl)-propionic acid (1). Received on a method [11], m.p. 66–67°C (compare Ref. [11] m.p. 66°C).

tert-butyl ester 3-(3', 5'-di-tert.butyl-4'-hydroxyphenyl)-propionic acid (2). To solution of 11.2 g (0.1 mol) Bu^{t}OK in 200 ml Bu^{t}OH have added 29.7 g (~ 0.1 mol) chloride 3-(3', 5'-di-tert.butyl-4'-hydroxyphenyl)-propionic acid, a reaction mixture maintained at boiling point ~2.5 h, distilled off solvent, the residual extracted ether. From a mother solution evolved 29 g (87 %) 2, m.p.7071°C (from hexane). ^{1}H NMR spectrum (CDCl$_3$, δ): 1.44 (s, 18 H, Bu^t); 1.56 (s, 9 H, COOBu^t); 2.62 (t, 2 H, CH$_2$CH$_2$CO); 2.89 (t, 2 H, CH$_2$CO); 5.1 (s, 1 H, OH); 7.1 (s, 2 H, Ar). The IR-spectrum, ν/cm^{-1}: 3643 (OH); 2957 (CH); 1717 (COO); 1433; 1366; 1135 (C-O-C); It is found (%):75.38; H, 10.43. C$_{21}$H$_{34}$O$_3$. It is calculated (%): C, 75.40; H, 10.25.

[2,2-(methylcarboxymethyl)-2,6-di-tert-butylmethylenchinone] (1b). Of 0.9 g (~3.0 mmol) 1 and 4 g (~17 mmol) Ag$_2$O in 30 ml of hexane maintained an admixture ~1 h at 20°C. After branch Ag$_2$O solvent distilled. From the residual with fractional crystalisation with the subsequent chromatographing on Al$_2$O$_3$ have evolved 0.42 g (49 %) 1b; m.p. 161°C (compare ref.[12] m.p.159–161°C). ^{1}H NMR spectrum (CDCl$_3$, δ): 1.24 (s, 18 H, Bu^t, Ar); 1.32 (s, 18 H, Bu^t, Ar); 1.84 (s, 6 H, COOCH$_3$); 4.19 (d, 2 H, CH = CHCO, J = 9.1 Hz); 5.88 (d, 2 H, CH = CHCO, J = 9.1 Hz); 6.73 (s, 2 H, Ar); 7.29 (s, 2 H, Ar).

[2,2-(tert.butylcarboxymethyl)-2,6-di-tert.butylmethylenchinone]
(2b). Received similarly from 1 g (~ 3.0 mmol) 2 and 4 g (~17 mmol)
Ag_2O. A yield 2b 0.13 g (~14 %); m.p. 193–194°C (from hexane). ^1H
NMR spectrum ($CDCl_3$, δ): 1.25 (s, 18 H, Bu^t, Ar); 1.33 (s, 18 H, Bu^t,
Ar); 1.43 (s, 18 H, $COOBu^t$); 4.20 (d, 2 H, CH=CHCO, J = 9.2 Hz); 5.90
(d, 2 H, CH=CHCO, J = 9.2 Hz); 6.75 (s, 2 H, Ar); 7.31 (s, 2 H, Ar). The
UV-spectrum (λ max, hexane) 292 (lg ε =4.02) and 315 (lg ε =4.34); the
IR-spectrum, ν / cm^{-1}: 1735 ($COOBu^t$), 1655 (C=O), 1640 (C=C). It is
found (%):75.98; H, 9.53. $C_{42}H_{62}O_6$. It is calculated (%): C, 76.13; H, 9.39.
The initiated oxidation of *iso*-propyl benzene (ArH) oxygen.

A. In the reactionary vessel of gasometric installation seated 20 ml of
   solution 2.6–3 × 10^{-5} mol·l^{-1} compound 1 (or 2), 5 × 10^{-3} mol·l^{-1}
   azodiisobutyronitrile (ADBN) in iso-propylbenzene (ArH), at
   50°C and filled O2 at atmospheric pressure. Kinetic measurements
   made on a method [2].

B. In container on 100 ml seated 50 ml of the solution keeping
   2×10–2 mol·l–1 compound 2, 2×10–2 mol·l–1 ADBN in ArH, and
   at 50°C maintained in O_2 at atmospheric pressure within 50 min
   Solvent evaporated, to the residual have added 10 ml of hexane,
   heat to 50°C, further cooled to 0–3°C, filtrated, the residual dissolved in $CDCl_3$ and analyzed a NMR method ^1H. ^1H NMR spectrum ($CDCl_3$, δ): 0.92 (s, 6 H, CH_3); 1.39 (s, 18 H, Bu^t, Ar); 1.51
   (s, 18 H, $COOBu^t$); 2.74 (t, 1 H, CH_2-CH-Ar, J = 7.4 Hz); 2.93
   (d, 2 H, CH_2-CHCO, J = 7.4 Hz); 5.22 (s, 1 H, OH); 7.1 m, 7
   H, Ar) corresponds to *tert-butyl* ester 3-(3', 5'-*di-tert-butyl*-4'-hy-
   droxyphenyl)-3 (2-'peroxy-2', 2'-dimethylbenzile)-propionic acid.
   At chromatographed on Al_2O_3 there is a decomposing.

2.3   DISCUSSION OF RESULTS

The mechanism of considerably-chain oxidation of hydrocarbons is known
and based on kinetics of unit steps [2]. In practice of research of laws use
the initiated oxidation with constant speed of formation alkyl radicals. At
the initiated oxidation of iso-propyl benzene (ArH) at presence azodiiso-

butyronitrile, oxygen and a reaction antioxidant proceed according to the schema 1.

2.3.1   SCHEMA 1

ADBN – Azodiisovutyronitrile; InOH – Antioxidant, (1,2)

The *iso*-propyl benzene radical (Ar ˙) is formed in reaction azodiiso-butyronitrile (ADBN) with ArH (1) of which in the presence of oxygen it is formed peroxy radical ArOO ˙(2). Continuation of considerably-chain oxidation proceeds in reaction ArOO ˙ with ArH (3) with a reaction constant k_2 (according to Ref. [4] $k_2 = 1.75$ l·mol^{-1} c^{-1}); inhibition – in reaction ArOO ˙ with antioxidant InOH (4) with a reaction constant k_7. Specific reaction rate value (k_7) defined from dependence $\Delta [O_2]/[ArH] = -k_2/k_7 \cdot \ln (1-t \times \tau^{-1})$ where, [ArH] – concentration ArH (7.18 mol·l^{-1}), $\Delta [O_2]$ – the quantity turned sour-sort, entered reaction with ArH. The season of inhibition of oxidation ArH oxygen with antioxidant participation (τ) and value of stopping of chain (f) at constant speed initiation ($W_i = 1.5 \times 10^{-8}$ mol l^{-1}s^{-1}) defined a gasometric method (Fig. 1).

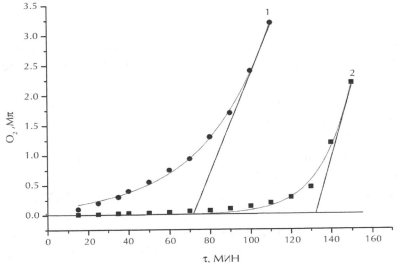

FIGURE 1   *(Continued)*

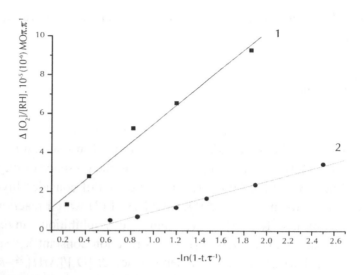

FIGURE 1 Kinetic dependences of the expense of oxygen in reaction of initiated oxidation ArH in the presence of antioxidants 1 and 2 and them is at 50°C. фозы. $[1]_o = 3 \times 10^{-5}$; $[2]_o = 2.6 \times 10^{-5}$ mol·l⁻¹. For 2 size $\Delta [O_2] / [RH] \times 10^6$ (on axis Y).

The methyl ester 3-(3', 5'-*di-tert.*butyl-4'-hydroxyphenyl)-propionic acid (1) is known as an antioxidant [12] and is characterized by values $k_7 = 2.3 \times 10^4$ l·mol-1·s⁻¹, f = 2, and this data is confirmed by the present work. However as a result of inhibiting oxidation at presence tert.butyl ester 3-(3', 5'-di-tert.butyl-4 '-hydroxyphenyl)-propionic acid (2) reaction proceeds with a specific reaction rate $k_7 = 3.0 \times 10^6$ l.mol⁻¹·s⁻¹, coefficient of stopping of chain f = 5.

Butyl bunches in ortho positions of a molecule of phenols shield a hydroxyl group, and bridging of nuclear orbitals of atoms of hydrogen of the next bunches is characterized by energy of steric interaction (steric energy). Similar influence is rendered *tert.*butyl by the substituent at carboxyl group oxygen, shielding the next atom of carbon. These can explain distinctions in reactivity of two homologues – derivatives 3-(3', 5'-*di-tert.*butyl-4'-hydroxyphenyl) – propionic acid, different among themselves alkyl substituents: methyl or *tert.*butyl at atom of oxygen of a carboxyl group. It is known through Ref. [1] that in inhibiting oxidation in the presence of an antioxidant of a class of phenols are formed phenolic radicals, which properties can affect on process inhibiting oxidations. Also it is known through Ref. [13] that at one-electronic oxidation of compound 1 reaction proceeds according to the Schema 2.

2.3.2 SCHEMA 2

Taking into account this data establishing the fact of formation phenoling radicals and products of their transformations under the Schema 2, oxidation of compounds 1 and 2 in relative conditions is made. At oxidation of an antioxidant 1 Ag_2O at 20°C within 1 h compound 1b (R=Me) with a yield of 49 % is formed, at oxidation 2 in simulated condition - yield 2b 14 %. Last result is interpreted as consequence of influence tert.butyl substituent at carboxyl group oxygen that leads to yield decrease methylenchinone 2b in comparison with 1b (14% against 49%). In work recombination possibility phenolic radical 1a with radical ArOO˙ is considered. Peroxide formation 3a is confirmed by the yielded NMR ^1H a spectrum on presence of signals from protons CH_2 (δ = 2.93) and CH (δ= 2.74) Bunches in reactionary mass after initiated oxidation ArH in the presence of compound 2 (the schema 3). Examples of reactions phenolic radicals with peroxy radicals are known, and products of their recombination's break up with allocation of oxygen [14, 15].

The peroxide structure 3a is based on the data of calculations by a quantum-chemical method in approach PM6, starting with energy formation of structures 3a–3d. To a minimum of energy of formation (H_f° =-804.9 kJ.mol^{-1}) there corresponds structure 3. For frames: H_f° –692.9, 3b –710.7, 3c –717.1, 3d kJ.mol^{-1}.

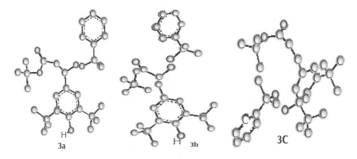

3a   3b   3C

FIGURE 2   *(Continued)*

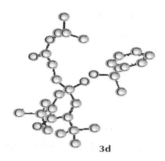

FIGURE 2   Structures recombination products phenolic radical 2a with ArOO.

Frames 3a and 3b—are analogues of phenols structure, frame 3c and 3d—analogues cyclohexadienons structure . From calculation of frames 3a and 3b in program MM2 follows that energy of steric interaction (Hs) in 3a is equal 76.9 kJ.mol⁻¹, in 3b > 1000 kJ.mol⁻¹ that, apparently, is a consequence influence of a steric factor on a direction of recombination of radicals from an antioxidant 2.

Convertibility of reaction (4) follows from quantum-chemical calculations in approach PM6 of the software package "мопак , 2009". For compounds 1 and 2 have calculated energy of formation (E_f^o), enthalpies (H°), entropies (S_f^o), and energy of a cleavage of communication O–H bond (D_{OH}) in antioxidants 1 and 2 (Table 1) have defined from the formula [16]:

$$D_{(OH)} = -E_f^o (\text{In} \cdot) + H_f^o (\text{H}) - (-E_f^o \text{InH}); \; H_f^o (\text{H}) = 218.0 \text{ kJ.mol}^{-1}$$

TABLE 1   Energy of formation, enthalpy, entropy and energy O–H of communications of compounds 1 and 2 and them phenoxyl radicals 1a and 2, 298 K.

Compounds	-Efo	Hfo	Sfo	DOH
		:kJ.mol–1	J/K/mol–1	kJ.mol–1
1	667.4	16.2	170.4	310.5
1a	569.4	16.1	174.2	313.4
2	769.9	19.1	191.5	
2a	674.0	19.0	198.1	

Proceeding from the settlement data of enthalpy and entropy of the compounds participating in reactions of antioxidants 1 and 2 with ArOO ˙, on the equation of Gibbs have calculated thermodynamic equilibrium constants K (Table 2).

TABLE 2 Change of thermodynamic functions and energy of Gibbs in reaction of compounds 1 and 2 (InH) with ArOO˙, 298 K.

Compounds	ΔH°_f kJ·mol⁻¹	ΔS°_f kJ·K·mol⁻¹	ΔG°_f kJ·mol⁻¹	Δ^x
1	+ 200.03	. + 0.94	− 80.68	1.03
2	+ 283.14	+ 3.79	− 846.28	1.41

ˣEnergy of formation ArOO, $E_f^{\circ} = -46.4$ kJ.mol⁻¹; ArOOH, $E_f^{\circ} = -50.3$ kJ.mol⁻¹.

From this data follows that in the isolated conditions the equilibrium constant is close to value equal 1, that confirms balance possibility in reaction (4). In non-polar solvent balance is displaced towards formation ArOOH and InO ˙ [17].

2.4   CONCLUSION

Thus, experimental and settlement yielded inhibiting oxidations ArH in the presence of antioxidants 1 and 2 allow to interpret results taking into account influence of structure of an antioxidant. In this case, formed phenolic radical (In·) interacts with ArOO ˙ that leads to augmentation of speed of inhibition and augmentation of time of an induction period of oxidation ArH.

KEYWORDS

- Azodiisobutyronitrile
- Gasometric method
- Hartrii-Fock
- Iso-propyl benzene
- Phenoling radicals

REFERENCES

1. Denisov, E. T. *Oxidation and destruction of carbocyclic polymers*, Chem, Leningrad, **1990**, *192* (Russian).
2. Emanuel, N. M.; Denisov, E. T.; Majzus, K. *Chain reactions of oxidation of hydrocarbons in a fluid phase*, Science, Moscow, **1965**, *375* (Russian).
3. Roginsky, V. A. *Phenolic antioxydantes :Reaction ability and efficacy*, Science, Moscow, **1988**, *247* (Russian).
4. Djubchenko, O. I.; Nikulina, B. V.; Terah, E. I.; Prosenko, A. E.; Grigoriev, A. *Izv. AH, Ser.him.*, **2007**, *1107* [*Russ. Chem. Bull., Int. Ed.*, **2007**, *56,* 295].
5. Roginsky, V. A. *Kinetics and catalysis*, **1990**, *546* (Russian).
6. Beljakov, V. A.; Shanin, E. L.; Roginsky, V. A.; Miller, V. B.; *Izv. AH the USSR, Ser. him.* **1975**, *2685.*
7. Denisov, E. T.; *J. Phys. Chim.*, **1995**, *69,* 623 (Russian).
8. Hursan, S. L. The *Receiving tank of lectures on VIII the international conference "Bioantioxy-dant"*, Moscow, **2010**, *195* (Russian).
9. Berdyshev, D. V.; Glazunov, V. P.; Novikov, V. L.; Izv. The Russian Academy of Sciences, Ser. Khim., **2007**, *400* [*Russ. Chem. Bull., Int. Ed.*, **2007**, *56, 295*].
10. Stewart, J. J. P. *J. Mol. Mod.*, **2007**, *13,* 1173.
11. Volodkin, A. A.; Zaikov, G. E.; Izv. The Russian Academy of Sciences, Ser.Khim., **2002**, *2031* [*Russ. Chem. Bull., Int. Ed.*, **2005**, *55*].
12. Storozhok, N. M.; Perevozkina, M. G.; Nikiforov, G. A.; Izv. The Russian Academy of Sciences, Ser. Khim., **2005**, *323* [*Russ. Chem. Bull., Int. Ed.*, **2005**, *54, 295*].
13. Kudinova, L. I.; Volodkin, A. A.; Ershov, V. V.; Izv. The Russian Academy of Sciences, Ser.Khim., **1978**, *2797* (Russian).
14. Ershov, V. V.; Nikiforov, G. A.; Volodkin, A. A.; Chemistry of Steric Phenols, Chem, Moscow, **1972**, *151* (Russian).
15. Kovarova-Lerchova, J.; Pospisil, J. *Eur. Polymer*, **1977**, *3,* 975.
16. Denisov, E. T. *Methods of definition of dissociation energy O-H of communication in Phenolums, the Receiving tank of lectures on VIII the international conference "Bioantioxidant"*, Moscow, **2010**, *50* (Russian).
17. Luzhkov, V. B. Chem. *Phys.*, 2005, 314, 211.

CHAPTER 3

ASSESSMENT OF THE POTENTIAL OF ENHANCED OIL RECOVERY FROM RESERVOIRS WITH HIGH WATER CONTENT USING THE HEAT OF NITRATE OXIDATION REACTIONS AND IN SITU HYDROCARBON OXIDATION

E. N. ALEKSANDROV, P. E. ALEKSANDROV, N. M. KUZNETSOV, A. L. PETROV, and V. YU. LIDZHI-GORYAEV

CONTENTS

3.1 INTRODUCTION

A technique is described for enhanced oil recovery using the heat and gas produced by ammonium and organic decomposition reactions taking place in wellbores and reservoirs. A comparative analysis is presented of the contributions to the process from decomposition reactions of inorganic (ammonium) and organic (monoethanolamine, MEAN, $HO-CH_2-CH_2-NH_3^+NO_3^-$) nitrates. According to calculations, the amount of gas produced is higher and the heat of reaction is lower for high-temperature reaction of MEAN as compared to ammonium nitrate, and the opposite is true for low-temperature reactions. Field tests and calculations have demonstrated that binary mixture technology can provide an alternative to well-known technologies of enhanced oil recovery. Three methods are proposed for stimulating reservoirs via absorption of gas by oil and ensuing reduction of reservoir fluid viscosity and density. Heterogeneous catalytic processes that occur in heated reservoirs must be analyzed and taken account of as promising research directions in thermochemical oil recovery methods.

To improve oil recovery, reservoir pressure is increased by water injection into the reservoir [1]. The adverse mobility ratio between oil and water results in early water breakthrough and low displacement efficiency. The average water content in reservoirs across Russia is 50%, and the average water cut from producing wells in Russia is 84%. When the percentage of oil in the produced fluid drops to 10%, production is usually terminated. In such cases, the amount of remaining oil in place is comparable to the total amount of oil recovered.

Owing to the wide use of steam-assisted gravity drainage (SAGD) in heavy-oil production, Canada became one of the world's top ten hydrocarbon-producing countries. Heavy-oil recovery by steam injection requires two to five tonnes of steam per tonne of recovered oil [2-4]. After 20% of the heavy oil in place has been recovered from a reservoir, most of the steam energy is wasted to heating the water contained therein, and the profitability of production decreases by a factor of at least 2 [3, 4].

When using a thermochemical technology based on binary mixture (BM) reaction, the increase in reservoir water content is negligible compared to SAGD and water-injection technologies. Reservoir heating by binary mixture reaction is achieved by injecting solutions of chemicals, ammonium nitrate or MEAN and a nitrate decomposition initiator, into a well through separate passages [3, 4]. The hot, high-pressure gas released at the bottomhole enters the reservoir.

The main drawback of the thermochemistry currently used to enhance oil recovery is a lack of control over downhole reaction. Since ammonium nitrate decomposition by the reaction $NH_4NO_3 \rightarrow N_2 + 2H_2O + 0,5O_2 + Q_1$ produces heat, nitrogen, and oxygen, Rostechnadzor (Russian Technical Supervision Service) has restricted the amount of ammonium nitrate injected to one tonne per well since it would pose an explosion hazard when used in large amounts.

Control of thermochemical processes under downhole and reservoir conditions is key to both work safety and optimization of enhanced oil recovery methods. In 2010, Emmanuel Institute of Biochemical Physics presented to Rostechnadzor a mobile laboratory that controls reaction under downhole conditions and ensures safe injection of large amounts of nitrates into the reservoir. Rostechnadzor approved experimental injection of an unrestricted amount of nitrates into boreholes under the requirement of two levels of safety control [5].

In October/November 2011, more than 26 tonnes of BM chemicals were injected into wells nos. 1242 and 3003 of a Permo–Carboniferous reservoir in the Usinsk oil field (Lukoil–Komi LLC).

3.2 EXPERIMENTAL

3.2.1 *CONTROL OF BM REACTION TEMPERATURE UNDER DOWN HOLE AND RESERVOIR CONDITIONS*

Aqueous solutions of ammonium nitrate and a decomposition initiator (sodium nitrite) were injected through separate passages. The aqueous solutions injected into well no. 1242 contained 4 tonnes of the nitrate and 2 tonnes of the initiator. The performance of temperature and pressure sensors was checked in different pH solutions while a nitrate decomposition reaction was taking place in the borehole. A stable operation range was determined for the sensors (4 < pH < 8).

Since no data could be found on controlled decomposition of an explosive compound (nitrate) under oil-field conditions, we describe in some detail the "tubing-inside-tubing" injection scheme and the procedures used to control BM reaction at the bottomhole during injection. More than 20 tonnes of aqueous solutions (with 66 wt% NH_4NO_3 and 50 wt% $NaNO_2$, respectively) were injected into well no. 3003 in the Usinsk oil field. To do this, a 73-mm diameter tubing string was run through 168-mm casing perforated in the reservoir zone to the reservoir base at a depth of 1401m, and a 45-mm diameter tubing string was run through the 73-mm tubing to a depth of 1391m. A packer was set in the annulus between the casing and the 73-mm tubing at a depth of 1301 m. The ammonium nitrate solution was injected through the annulus between the 73- and 45-mm tubing strings, and the sodium nitrate solution was injected through the 45-mm tubing. Inside the 45-mm tubing, there was a cable with temperature sensors at depths of 1301 m (packer setting depth) and 1401m and a pressure sensor at a depth of 1301m.

Injection safety was ensured as follows: (1) a 67 wt % nitrate solution was injected in portions of no more than 1 m^3 alternately with water slugs (50 to 150 liters); (2) reaction parameters at 1401 m were controlled by varying the flows of the reaction initiator and the additional water used to quench the reaction. If reaction rate exceeded a certain limit, the injection of chemicals would be stopped. The reaction was initiated near the bottomhole end of the 45-mm tubing string (initiator outlet).

Fig. 1 shows temperature (dashed curve) and pressure (dash–dot curve) near the packer at a depth of 1301 m and temperature in the reaction zone at 1401 m (solid curve). In both wells, nos. 1242 and 3003, the injection process consisted of several cycles where the initiator, the nitrate, and water were injected successively.

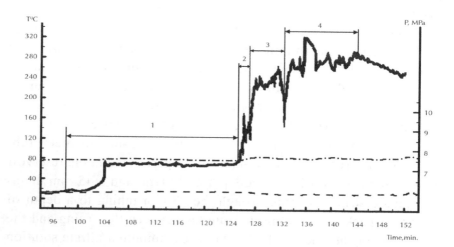

FIGURE 1   Treatment of the reservoir at the bottom hole of well no. 3003 in the Usinsk oil field with BM reaction products. Temperature histories in the reaction zone (solid curve); temperature inside the tubing near the packer (dashed curve); pressure inside the tubing near the packer (dash–dot curve). Cycles of chemical injection are numbered by *1, 2, 3, 4.*

The injection of an ammonium-nitrate-based BM into well no. 3003 was performed in four cycles between the 98th and 144th minutes (see Fig. 1). During the 46-min interval, 7.5 m³ of solutions containing 5 tonnes of NH_4NO_3 and 2.3 tonnes of $NaNO_2$ (initiator) were injected.

After the first portions of chemicals (1 m³ of nitrate + 1 m³ of initiator + 0.05 m³ of water) were injected at the 98th minute, temperature at the bottomhole increased from 14 to 76°C at the 104th minute. In other words, the characteristic reaction time was $\tau = 6$ min for an initial temperature of 14°C. The second injection of BM solutions (2 m³) and water (0.05 m³) was performed at the 125th minute. The temperature rose to 170°C at the 126th minute; i.e., $\tau = 2$ min for an initial temperature of 76°C. Having reached a local maximum, the temperature dropped by 40 degrees the next minute as the water slug (0.05 m³) passed through the reaction zone.

The next injection (4 m³ of BM + 0.05 m³ water slug) started at the 127th minute. A temperature rise to 245°C was observed. Further injection of 1.5 m³ of BM started at the 130th minute, followed by injection of a large water slug (120 kg) at the 133rd minute. At the 134th minute, the temperature dropped by 80 degrees because the reaction zone was cooled by this amount of water.

To test the effectiveness of the initiator at high temperature, 100 liters of the initiator solution were injected between the 135th and 136th minutes simultaneously with the nitrate. As a result, the reaction zone temperature increased from 250 to 325°C as the chemicals were injected during the 136th minute. The temperature of 325°C was set as a limit at which initiator injection may accelerate the reaction to a hazardous rate ($\tau < 1$ min).

It is clear from Fig. 1 that, while the temperature in the reaction zone at a depth of 1401 m varied between 250 and 326°C, the temperature near the packer at 1301 m remained within the range of 18 to 20°C, and the downhole pressure during the interval between the 98th and 154th minutes did not exceed 4 atm. This can be attributed to the high injectivity of a reservoir containing numerous cracks.

Thus, it was demonstrated for the first time that both high extent and safety of controlled downhole reaction can be achieved simultaneously.

The static fluid level in well no. 3003 was at a depth of 410 m during the injection. To measure the speed of rise of the fluid level in the well, the device that measured temperature and pressure at the packer setting depth was moved to a depth of 400 m.

At the 155th minute, the well was found to be poorly sealed near the wellhead. To complete the injection schedule after a sealing failure, the final portion of approximately 9 m³ of a solution containing approximately 6 tonnes of NH_4NO_3 was injected, and the nitrate solution remaining in the wellbore was removed by injecting 1 m³ of fresh water.

After the injection of chemicals, the well must remain shut for a period of several hours. This is required to prevent the packer from overheating due a gas lift effect similar to that previously observed in well no. 169 in the Kurbatovskoe oil field [4]. Gas lift occurs when a fluid mixed with a gas flows up the wellbore. Most of the heat and gas produced by reaction should be absorbed by the reservoir in 3 to 6 hours. In two hours after the final injection of chemicals, a rise in fluid level from a depth of 410 m to the surface was detected, while the fluid temperature near the wellhead increased from 0 to 48°C. Apparently, the reaction that produced a mass of gas similar to the mass of injected nitrate gave rise to a *thermochemical gas lift* effect [4], which proved to be instrumental in bringing reservoir fluid to the surface. The thermochemical gas lift driven by BM reaction was discovered and examined on a qualitative level after reaction had been initiated by injecting 1.5 tonnes of BM chemicals into well no. 169 in the Kurbatovskoe field [4]. The more powerful thermochemical gas lift created after over 20 tonnes of BM chemicals were injected into well no. 3003 in the Usinsk field lends itself to a quantitative analysis.

To evaluate the pumping power for the thermochemical gas lift created in well no. 3003 in the Usinsk field, we note that a fluid column with a weight of 20 tonnes was gas-lifted to a height of 410 m in two hours. Dividing the amount of work done (8,200,000 kilogram-meters) by the time of 7200 s, we obtain 11.2 kW. This power is sufficient for producing at least 200 tonnes of oil-containing fluid heated to 300–400°C per day.

According to the table, the incremental oil recovered from both wells by using the BM technique amounted to 1,397.831 t. This result was achieved by injecting 26 t of ammonium nitrate into the wells. The reservoir was stimulated by the 24 GJ of heat produced by the decomposition reaction Table 1.

TABLE 1 Results of BM technology tests in wells nos. 1242 and 3003 in the Usinsk oil field.

Well no.	Pump model	Month	Production restart	Number of days	Incremental oil, tonnes	Production, tonnes/day		
						Baseline	Average	Planned
		Nov 2011	Nov 9, 2011	22	127.996	0	5.818	
		Dec 2011		30.83	169.565		5.5	
1242	EVNT-25	Jan 2012		31	143,0		4.63	
		Feb 2012		29	143.26		4.94	
		Mar 2012		31	123.38		3.98	
		Total		143.83	707.731		4.92	8.5

TABLE 1 *(Continued)*

		Jan 2012	Jan 4, 2012	28	242.9	1.93	10.6	
3003	EVNT-25	Feb 2012		23	199.6		10.6	
		Mar 2012		30.75	247.6		9.98	
		Total		81.75	690.1		8.44	6.5

The incremental oil recovered from both wells by using the BM technique before April 2012 amounted to 1,398 tonnes. Before December 2012, approximately 2500 tonnes of incremental oil were produced. Calculations show that these results correspond to extent of reaction equal to 0.8. Uncontrolled reaction is conducted at lower temperatures with extent of reaction smaller by a factor of 2.

3.3   RESULTS AND DISCUSSION

3.3.1   ABSORPTION OF GAS BY OIL AND CREATION OF GAS LIFT

The incremental oil was recovered by means of decomposition of approximately 20 tonnes of ammonium nitrate injected into two wells. Nitrate decomposition and partial oxidation of oil generated approximately 18.5 GJ of heat and approximately 20 tonnes of gas, including 6.5 tonnes of nitrogen, 8.3 tonnes of water vapor, and 4 tonnes of oxygen. Hydrocarbon oxidation in the heated reservoir resulted in heat release and production of 5.2 tonnes of $H_2O + CO_2$. The gas lift effect created by controlled reaction can be instrumental in bringing hot fluid from reservoir to the surface.

The results of previous treatments of 20 wells with the products of uncontrolled downhole reaction, where the BM mass varied between 0.8 and 1.5 tonnes per well, were used to estimate the extent of reaction as 0.30 to 0.35. Optimization of high-temperature reaction in well no. 3003 resulted in extent of reaction equal to 0.8.

Apparently, the most important result is the relatively fast progress of non-explosive reaction of nitrate decomposition observed in a reservoir heated to 250–326°C. The characteristic reaction time τ in the heated reservoir is a few hours. Inside a 168-mm wellbore, $\tau < 1$ min at similar temperatures.

It should be recalled that the average reservoir pore size is less than 1 mm and the average crack width varies between 1 and 5 mm. A cubic meter of reservoir rock with porosity of 20% (taking into account cracks) and density of 2.5 t/m³ can accommodate no more than 200 liters of a nitrate solution containing approximately 134 kg of NH_4NO_3. The corresponding mass ratio of nitrate to water-containing rock is 1 : 19.

The probability of explosion for such a "mixture" with 95% inert content is close to zero. Therefore, a large reservoir area can be safely heated by nitrate decomposition as follows. Initially, approximately 100 tonnes of reservoir rock are heated to 200–250°C by reaction of 10 to 15 m³ of nitrate solution with initiator. After that, any amount of nitrate can be injected into the heated zone. The highest safety will be achieved in low-porosity reservoirs.

3.4   HEAT OF FORMATION FOR MEAN AND APPLICATION OF MEAN-BASED BINARY MIXTURES AT OIL PRODUCTION SITES

Even though widely used for reservoir heating, ammonium nitrate is far from being an optimal chemical agent because it poses an explosion hazard. When control of reaction under downhole and reservoir conditions is impossible or unreliable, organic nitrates should be used since their decomposition does not produce molecular oxygen and generates large amounts of product gases. Specifically, MEAN has passed proving ground testing and has been approved for application at production sites. MEAN has been successfully used to enhance oil recovery in well no. 21 in the Razumovskoe field (Saratov oblast) and in well no. 943 in the Shumovskoe field (Perm Krai) [2]. One setback for the use of MEAN is the lack of data for its heat of formation in the literature.

Organic nitrates differ from NH_4NO_3 in terms of heat and gas produced. A preliminary estimate shows that MEAN produces more gas and less heat at high temperatures; the opposite is true at low temperatures. To obtain accurate predictions, we performed the calculations of the heats of formation for MEAN and its decomposition products described below.

3.4.1 HEAT OF FORMATION FOR MEAN

According to measurements [8, 9], the heat of formation for MEAN is:

$$\Delta H_1 = -137.5 \pm 8 \text{ kcal/mol.} \tag{1}$$

It was pointed out in Ref. [8] that a standard deviation of ± 8 kcal/mol was obtained without allowing for systematic errors and the high hygroscopicity of the nitrates under study.

The significant uncertainty in measured results of [8] motivated an independent thermochemical calculation of ΔH_1 presented in [10]. According to Ref. [10],

$$\Delta H_1 = -149.3 \pm 8 \text{ kcal/mol} = -625 \pm 10 \text{ kJ/mol} \tag{2}$$

The calculated value of Eq. (2) agrees with that of Eq. (1) determined from experiment, with their uncertainty intervals overlapping between -139 and $-145 \pm$, where δ is an additional uncertainty of the result reported in [8]. Assuming that the actual value of ΔH_1 lies within the overlap interval, we obtain:

$$\Delta H_1 = -142 \pm 4 \text{ kcal/mol.}$$

3.4.2 DECOMPOSITION PRODUCTS OF MEAN

Consider now the temperature variation of organic-nitrate decomposition characteristics and the ensuing effect on the reservoir. At relatively low temperatures (roughly 100 to 500°C), MEAN decomposition occurs via the reaction:

$$HO-CH_2-CH_2-NH_3{}^+NO_3{}^- \rightarrow CO_2 + 2H_2O + N_2 + CH_4 + Q_A. \qquad (2a)$$

Reaction at higher temperatures follows a different path, producing syngas among other products [11, 12]:

$$HO-CH_2-CH_2-NH_3{}^+NO_3{}^- \rightarrow 2CO + 2H_2O + N_2 + 2H_2 + Q_B. \qquad (2b)$$

This obviously changes both heat of reaction and amount of product gas. The values of Q_A and Q_B are unknown but easy to calculate. In particular, Q_A is the difference between the heat of formation H_1 for MEAN and the total heat of formation for $CO_2 + 2H_2O + N_2 + CH_4$. Denoting the latter by $\Delta H_{A(g)}$, we have

$$Q_A = \Delta H_1 - \Delta H_{A(g)}; \qquad (3)$$

analogously,

$$Q_B = \Delta H_1 - \Delta H_{B(g)}, \qquad (4)$$

where $\Delta H_{B(g)}$ is the total heat of formation for the gaseous products of Eq. (2b). Both $\Delta H_{A(g)}$ and $\Delta H_{B(g)}$ are readily calculated by using the heats of formation found in Refs. [13–16] for the product components:

$$\Delta H_{A(g)} = -229.6 \text{ kcal/mol}, \qquad (5a)$$

$$\Delta H_{B(g)} = -169.9 \text{ kcal/mol}. \qquad (5b)$$

Note that the values in Eq. (5) are calculated per mole of MEAN. According to Eqs. (3)–(6)

$$\Delta H_{B(g)} - \Delta H_{A(g)} = Q_A - Q_B = 59.7 \text{ kcal/mol}. \qquad (6)$$

This difference is the heat of the reaction:

$$0.5CO_2 + 0.5CH_4 \rightarrow CO + H_2 + Q. \qquad (7)$$

Substituting the value of ΔH_1 evaluated above and that of Eq. (5a) into Eq. (3), we find that the heat of path (A) per kilogram of MEAN is 707 ± 80 kcal/kg.

Syngas composition not only varies with temperature and pressure but also depends on the source compound. Because no analysis of the decomposition products of MEAN has previously been reported, we present our calculations here.

The equilibrium pressures of the species involved in Eq. (7) satisfy the equation:

$$p_{CO} p_{H_2} / \sqrt{p_{CH_4} p_{CO_2}} = K_P \qquad (8)$$

where K_p is the equilibrium constant for Eq. (7). Values of the equilibrium constant for Eq. (7) at several temperatures can be found in Ref. [19]. The following equilibrium constants, whose values are given in Ref. [20], are sufficient for calculating K:

$$K_{P1} = p_C p_O / p_{CO}, \qquad (9)$$

$$K_{P2} = p_H^2 / p_{H2}, \qquad (10)$$

$$K_{P3} = p_C p_{O2} / p_{CO2}, \qquad (11)$$

$$K_{P4} = p_C p_H^4 / p_{CH4}, \qquad (12)$$

where p_C is the partial pressure of gaseous atomic carbon, p_O is the partial pressure of atomic oxygen, and so on (in atmospheres) [18]. Combining Eq. (8) with Eqs. (9)–(12), we obtain Eq. (13):

$$K_P = \sqrt{K_{P3} K_{P4}} / (K_{P1} K_{P2}) \qquad (13)$$

Table 2 presents the values of K_p calculated for several temperatures by using data on equilibrium constants in Eqs. (9)–(12) found in Ref. [18].

TABLE 2 Calculated values of K_p.

T, K	K_p, atm
400	3.85×10^{-10}
600	1.29×10^{-4}
800	0.082
900	0.739

The number of unknowns in Eq. (19) can be reduced by using the stoichiometric coefficients of the MEAN decomposition reaction and a material balance Eq. (14):

$$2p_{CO} + 2p_{CO_2} + 2p_{CH_4} = 3p_M \qquad (14)$$

where $p_M = n_M kT$ (n_M is the number of decomposed MEAN molecules per unit volume, k is the Boltzmann constant). Finally, we obtain a quadratic equation in a single variable p_{CO} instead of Eq. (8):

$$4p_{CO}^2 / (3p_M - 2p_{CO}) = K_P \qquad (15).$$

The product compositions in Eqs. (2a) and (2b) correspond to the limit cases of $p_{CO}/p_M = 0$ ($T \to 0$) and $p_{CO}/p_M = 1$ ($T \to \infty$), respectively. It should be noted that equilibrium product composition depends not only on temperature but also on the amount of decomposed MEAN (more precisely, on n_M), though much more weakly.

As an illustration of the temperature dependence of product composition, Table 3 presents several values of p_{CO}/p_M at $p_M = 1$ atm predicted by Eq. (15).

TABLE 3 Temperature dependence of p_{CO} / p_M at $p_M = 1$ atm.

T, K	400	600	800	900
p_{CO} / p_M	$1.7 \cdot 10^{-5}$	$9.8 \cdot 10^{-3}$	0.228	0.582

As p_M increases, the ratio p_{CO}/p_M decreases approximately as $p_M^{-0.5}$. According to Table 3, this implies a significant shift in equilibrium towards path (A) at $T < 800$ K and $p_M > 1$ atm. Analogous calculations using values of K_p given in Ref. [17, p. 72] lead to the same conclusion.

The shift in equilibrium is so dramatic that path (2a) remains predominant even under downhole conditions where nitrate decomposition products are in a dense fluid state. For example, the heat of reaction at a downhole temperature between 500 and 700 K should not change by more than 10% relative to standard conditions. Estimates for the energy budget under downhole conditions may be obtained without taking into account variations in heat of reaction and product composition with temperature and pressure.

These calculations agree with results of well treatments with organic-nitrate decomposition products in the Shumovskoe and Razumovskoe oil fields. Production from well no. 21 in the Razumovskoe field (Saratov oblast) and well no. 943 in the Shumovskoe field have increased by 430 and 210%, respectively.

3.5 ASSESSMENT OF THE POTENTIAL USE OF HIGH-TEMPERATURE BINARY MIXTURE REACTION WHEN EXPLOITING RESERVOIRS WITH HIGH WATER CONTENT

The deployment of a mobile laboratory at a production site will ensure work safety and control of reaction in wells and, to some extent, in the reservoir. As a result, progress can be expected in the following: (1) investigation and control of in situ oxidation and cracking processes, which requires practical work with both ammonium nitrate and organic nitrates that do not produce oxygen; (2) development of a technology for oil extraction from marginal reservoirs with high water content, which requires development and practical implementation of transition to a supercritical (SC) state of a water–oil mixture [6]; (3) development of an enhanced recovery technology where reservoir pressure is increased via absorption of gaseous products of exothermic reactions by reservoir fluid.

The BM technology has several advantages over SAGD. By controlling the rate of reaction of the heat- and gas-generating chemical, temperature and pressure can be raised almost without increasing water content. When the critical temperature $T_c = 374°C$ and pressure $P_c = 220$ atm are exceeded, water becomes supercritical and the number of hydrogen bonds per water molecule decreases [7].

A significant amount of lighter hydrocarbon fractions also become supercritical and miscible with water. Upon transition to an SC state, the resulting water–oil mixture becomes an easy-flowing homogeneous fluid with viscosity reduced by a factor of tens compared to that at average reservoir temperature (40–50°C) [4].

Hot fluid can be brought up to the surface in supercritical condition through thermally insulated tubing. As a result of implementation of this scheme in Russian oil fields, the average oil cut is expected to increase from 16 to 50%. Thus, the downward trend in profitability of production due to increase in water content can be reversed, and profitability improvement can be achieved by removing water from reservoirs (as part of water–oil mixture).

3.6 PRESSURE DRAINAGE AS AN ALTERNATIVE TO GRAVITY DRAINAGE AND THERMOCHEMICAL GAS LIFT EFFECT AS AN ALTERNATIVE TO ELECTRIC PUMP [19–22]

Let us now discuss the prerequisites and potential for enhanced oil recovery using reactive nitrate-based BM solutions injected into the upper well of the SAGD twin wells (Fig. 2). To inject BM solutions through separate passages, a dual-string completion is required. Pressure and temperature should be monitored continuously by using sensors placed in the reaction zone of the injection well and in the zone where hot reservoir fluid flows into the production well. Reaction in the reservoir surrounding the injection well results in a pressure rise. The product gas mixes with the reservoir fluid, reducing the fluid density and viscosity.

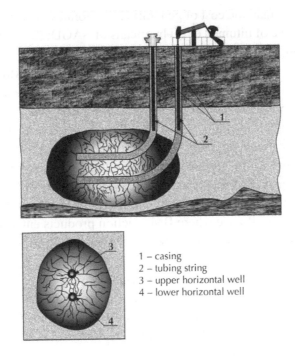

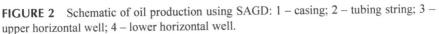

1 – casing
2 – tubing string
3 – upper horizontal well
4 – lower horizontal well

FIGURE 2   Schematic of oil production using SAGD: 1 – casing; 2 – tubing string; 3 – upper horizontal well; 4 – lower horizontal well.

We call the movement of reservoir fluid driven by gas pressure through pores and cracks from the injection to the production well *pressure drainage*. Unlike in steam-assisted gravity drainage, both intensity and direction of pressure drainage can be controlled. For instance, it can be directed upwards when BM reaction occurs in the lower horizontal well (see Fig. 2).

In view of these possibilities, continuous BM injection can be used in SAGD twin wells instead of steam injection, combined with extraction of heated gas-saturated fluid assisted by a gas lift effect. After some refinement, this mode of recovery may become economically advantageous even when applied in fields where steam generation facilities are available. This somewhat sweeping conclusion is substantiated by the following: (1) the cost of the chemicals consumed per tonne of incremental oil recovered from wells in the Usinsk oil field is lower than 160 rubles, which is lower than the cost of generating 2–5 tonnes of steam required to produce a tonne of bitumen or oil by means of SAGD; (2) the total time of BM injection and incremental oil recovery assisted by a thermochemical gas lift effect is much shorter than the time required to extract a similar amount of oil- or bitumen-containing fluid with pumps used in SAGD technology; (3) when applied to reservoirs with high water content under supercritical conditions (at $T > 647$ K), the gas lift effect should become an economical—and probably the only reliable—means for driving oil-containing fluid to the surface at a high rate. A necessary and sufficient condition for effective use of a refined BM technology is the availability of a reliable system for continuously monitoring pressure and temperature in the injection-well zone where BM reaction products enter the reservoir and in the production-well zone where the heated fluid flows out of the reservoir (Fig. 2).

Thermochemical gas lift acts as an engine using the energy generated by situ oxidation of hydrocarbons with oxygen produced by nitrate decomposition. Thermochemical gas lift can also be created via saturation of oil with the gas generated by burning a rocket propellant at the bottom-hole, as described in Ref. [21].

3.7   METHODS FOR INCREASING RESERVOIR PRESSURE BY MEANS OF EXOTHERMIC GAS-PRODUCING REACTIONS AS A MORE EFFICIENT ALTERNATIVE TO TECHNOLOGIES THAT INCREASE WATER CONTENT

Let us calculate the resources required to heat hundreds of thousands of tonnes of reservoir rock involving gas generation and creation of thermo-chemical gas lift effect and evaluate the expected oil recovery.

As an example, consider a reservoir zone with height of 10 m and horizontal area of 250,000 m^2, where wells are drilled for injecting aqueous solutions of chemicals (nitrate and decomposition initiator) and extracting (primarily gas-lifting) hot reservoir fluid. Suppose that the reservoir rock density is 2.5 t/m3, porosity is 20%, specific heat is 1 MJ/t·K, and reservoir temperature in the zone is to be raised by 200 K. A cubic meter of the rock contains 200 liters of oil with density of 0.9 t/m^3 and specific heat of

2.1 MJ/t·K. Then the total amount of oil in the reservoir zone is 4.5×10^5 tonnes.

Heating the reservoir rock and the oil contained therein will require approximately 500 MJ/m^3, which amounts to 1.25×10^6 GJ for the reservoir zone in question. Decomposition of one tonne of ammonium nitrate will generate approximately 1.6 GJ and 200 kg of oxygen. The ensuing oxidation of hydrocarbons will additionally generate up to 3.2 GJ. Therefore, 2.5×10^5 tonnes of nitrate are required to generate 1.25×10^6 GJ in the reservoir zone. Note that the mass of the gas produced by nitrate decomposition (nitrogen, oxygen, and water vapor) is equal to the mass of nitrate consumed.

Thus, 0.5 t of gas per tonne of heated oil must be generated in the reservoir and brought to the surface with recovered oil. This ratio of gas produced by chemical reaction to recovered oil (hereinafter called *thermochemical gas–oil ratio*) is sufficiently high to expect a rapid increase in well production rate even at an early stage of ammonium-nitrate injection into the reservoir, and at least 4.5×10^5 tonnes of oil will be gas-lifted to the surface. The amount of recoverable oil can be comparable to, or exceed, that of oil contained in the heated zone because additional oil will flow from the periphery towards the producing wells. Following Ref. [4], it is estimated that heating 4.5×10^5 tonnes of cold oil coming from the periphery will lower the reservoir zone temperature raised by 200 degrees by 50 degrees at most; i.e., reservoir heating can increase the amount of recoverable oil several times compared to the initial amount of oil contained in the zone. If the amount of chemicals specified above is pumped through injection wells for 1.5 years and thermochemical gas lift effect is created, then at least 820 tonnes per day on average can be extracted from production wells.

3.7.1 RESERVOIR HEATING WITH BOTTOMHOLE GAS GENERATOR BURNING HYDROCARBONS INJECTED FROM THE SURFACE

The required equipment should include pumps for injecting hydrocarbons and at least five compressors for delivering air at a pressure of 250 atm into the reservoir at a rate of approximately 180,000 m^3/h, which would suffice to oxidize 12 tonnes of hydrocarbons per hours. Heating the reservoir zone described above will take 1.5 years and require 3×10^4 tonnes of hydrocarbons to be burned, which should yield approximately 5×10^4 tonnes of CO_2. The mass of heated and recovered oil is estimated at approximately 5×10^5 tonnes. A thermochemical gas–oil ratio of 0.1 is sufficient for creating an effective gas lift effect.

3.7.2 OIL RECOVERY VIA DIRECT OXIDATION OF RESERVOIR HYDROCARBONS BY AIR OXYGEN

The most efficient method for reservoir heating and oil production via thermochemical gas lift effect should involve initial heating of a part of the reservoir by ammonium nitrate decomposition followed by injection of air into the reservoir [22]. The method proposed in Ref. [22] has been abandoned for 40 years precisely because of a lack of control over the process, which may well result in burning up as much as 90% of the oil in place. For this reason, the heating should be conducted while keeping in situ oxidation of hydrocarbons under control in order to prevent the formation of a combustion front.

To create a pressure of at least 250 atm, air compressors must provide a pumping rate of 36,000 m^3/h or higher. Injecting oxygen into the reservoir at a rate of 7200 m^3/h is sufficient for oxidizing 2.4 tonnes of hydrocarbons in situ per hour. Under these conditions, heating the reservoir zone described above will take 1.5 years. The mass of oxidized hydrocarbons will amount to 3×10^4, or 6.7% of the total oil in the heated reservoir zone. When the CO_2-to-fluid mass ratio is 1 : 10, a gas lift effect can be created. For the gas lift to be effective, it may be required to alternate between air injection into the heated reservoir and injection of nitrate leading to a higher gas/fluid ratio (1 : 2).

3.8 CONCLUSION

As noted above, one prerequisite for successful use of thermochemical technologies is the availability of a reliable system for continuously monitoring pressure and temperature under downhole and reservoir conditions. In contrast to SAGD, BM injection can be conducted by using a readily deployable treatment system and conventional production equipment. The injection of an amount of BM solutions sufficient to produce 2500 tonnes of incremental oil from wells nos. 1242 and 3003 took approximately 10 hours, including the time for deploying and testing the treatment system, whereas the injection of several thousands of steam required to produce 2500 tonnes of incremental oil would last at least a month.

Tests of the refined BM technology combined with subsequent calculations and estimation have shown that this technology offers a promising alternative to SAGD and water injection for increasing reservoir pressure.

Successful solution of the key problem of monitoring and control of thermochemistry under downhole conditions makes it possible to suggest promising research directions. These include investigation and control of in situ oxidation and cracking of hydrocarbons, which requires practical work with organic nitrates whose decomposition does not produce oxygen; exploring the possibility of exploitation of marginal reservoirs with high water content by methods using SC states of water–oil mixtures [7]; development of a technology for increasing reservoir pressure via absorption of heat and gas by oil with a view to extracting the remaining oil in place as part of a water–oil mixture. As a result, the rate of increase in water content must drop by an order of magnitude, while daily production assisted by a gas lift effect must increase several times. It can also be expected that sweeping efficiency will increase while operating costs per tonne of incremental oil will be reduced.

If the projected BM technology development proves successful, the expected results will improve the overall efficiency of industrial power generation based on fossil hydrocarbons, because increasing water content of reservoirs is an urgent problem in petroleum engineering.

KEYWORDS

- Binary mixture
- MEAN
- Steam-assisted gravity drainage
- Thermochemical gas lift effect

REFERENCES

1. Griguletskii, V. G. *Tekhnologii TEK,* April **2007**, 10 (in Russian).
2. Butler, R. M. *Thermal Recovery of Oil and Bitumen, Prentice-Hall*: Englewood Cliffs, **1991**.
3. Aleksandrov, E. N.; Varfolomeev, S. D.; Lidzhi-Goryaev, V. Yu.; Petrov, A. L. Enhanced Oil Recovery Using Binary Mixture (BM) Reaction Products As an Alternative to Increasing Reservoir Water Content, *Tochka Opory, 15 (159),* 4 December **2012** (in Russian).
4. Aleksandrov, E. N.; Kuznetsov, N. M. *Karotazhnik, (4),* 113–127, **2007** (in Russian).
5. Program and Technique for Operations to Be Performed in Testing Binary Mixtures (BMs) in a Permo–Carboniferous Reservoir of the Usinsk Oil Field, approved November 15, **2010** by Pechora Administration of the Russian Technical Supervision Service Agency, Resolution *25*, ID-19542–2010 (in Russian).
6. Merzhanov, A. G.; Lunin, V. V.; Alexandrov, E. N.; Petrov, A. L.; Lidzhi-Goryaev, V. Yu. High-Temperature Enhanced Oil Recovery, *Nauka i Tekhnologii v Promyshlennosti,* **2010**, *2*, 1–6, (in Russian).
7. Ved', O. V. et al., *Sverkhkriticheskie Flyuidy: Teoriya i Praktika,* **2007**, *2(2),* 55–69 (in Russian).
8. Cottrell, T. L.; Gill, J. E. *J. Chem. Soc.,* 1798–1800, **1951**.
9. Karapet'yants, M. Kh.; Karapet'yants, M. L. *Osnovnye termodinamicheskie konstanty neorganicheskikh i organicheskikh veshchestv* (Basic Thermodynamic Constants for Inorganic and Organic Compounds), Moscow: Khimiya, **1968** (in Russian).
10. Aleksandrov, E. N.; Aleksandrov, P. E.; Kuznetsov, N. M.; Lunin, V. V.; Lemenovskii, D. A.; Rafikov, R. S.; Chertenkov, M. V.; Shiryaev, P. A.; Petrov, A. L.; Lidzhi-Goryaev, V. Yu. *Sverkhkriticheskie Flyuidy: Teoriya i Praktika,* **2012**, *7(3),* 56–66 (in Russian).
11. Sheldon, R. A. *Chemicals from Synthesis Gas:Catalytic Reactions of CO and H_2,* Dordrecht: Reidel, **1983**.
12. Karakhanov, E. A. *Sorosovskii Obraz. Zh.,* **1997**, *3*, 69–74, (in Russian).
13. Dean, J. A. *Lange's Handbook of Chem.,* 15th ed., New York: McGraw-Hill, **1999**.
14. Vedeneev, V. I. et al., *Energii razryva khimicheskikh szyazei, potentsialy ionizatsii i srodstvo k elektronu: Spravochnik* (Bond-Dissociation Energies, Ionization Poten-

tials, and Electron Affinity: A Handbook), Moscow: Izd-vo Akad. Nauk SSSR, **1962** (in Russian).

15. *Kratkii spravochnik fiziko-khimicheskikh velichin* (Concise Handbook of Physical and Chemical Data), Mishchenko, K. P.; Ravdel, A. A., Eds., Leningrad: Khimiya, **1967** (in Russian).

16. Magaril, R. Z. *Mekhanizm i kinetika gomogennykh termicheskikh prevrashchenii uglevodorodov* (Mechanism and Kinetics of Homogeneous Thermal Transformations of Hydrocarbons), Moscow: Khimiya, **1970** (in Russian).

17. Mel'nikov, E. Ya. *Spravochnik azotchika* (Nitrogen Engineering Handbook), *1*, Moscow: Khimiya, **1967** (in Russian).

18. Gurvich, L. V. et al., *Termodinamicheskie svoistva individual'nykh veshchestv* (Thermodynamic Properties of Individual Substances), *1*, part 2, *Tablitsy termodinamicheskikh svoistv* (Tables of Thermodynamic Properties), 3rd Ed., Moscow: Nauka, **1978** (in Russian).

19. Aleksandrov, E. N.; Lemenovskii, D. A.; Koller, Z.; Gas Evolving Oil Viscosity Diminishing Compositions for Stimulating the Productive Layer of an Oil Reservoir, Patent WO 2010/043239, April 22, **2010**.

20. Aleksandrov, E. N.; Lemenovskii, D. A.; Koller, Z., Method and Apparatus for Thermally Treating an Oil Reservoir, Patent WO 2012/025150, March 1, **2012**.

21. Papusha, A. I. Method for Thermochemical for Neutralizing Highly Toxic Substances, RF Patent 2 240 850, January 30, **2004**.

22. Bokserman, A. A. Results and Prospects of Application of Thermal Reservoir Treatment Methods, in *Thermal Reservoir Treatment Methods,* Moscow: VNIIOENG, **1971**, 10 (in Russian).

APPENDIX

ENHANCED OIL RECOVERY USING BINARY MIXTURE (BM) REACTION PRODUCTS AS AN ALTERNATIVE TO INCREASING RESERVOIR WATER CONTENT

E. N. ALEKSANDROV, P. E. ALEKSANDROV, N. M. KUZNETSOV, V. YU. LIDZHI-GORYAEV, and A. L. PETROV

CONTENTS

A.1   INTRODUCTION

The potential of enhanced oil recovery using a refined binary mixture (BM) technology is described. A system for reaction monitoring and control developed by the authors is used to optimize heat and gas generations under down hole conditions and increase production by a factor of 3 to 6. The possibility is outlined of replacing the water injection currently used to increase reservoir pressure with gas generation by BM reaction. Unlike water, the product gas mixes with the reservoir fluid, reducing its viscosity and facilitating fluid flow up the wellbore. It is also found that heterogeneous catalytic reactions contribute to the creation of a thermochemical gas lift effect deemed instrumental in bringing reservoir fluid to the surface. High-temperature reaction is considered a promising regime under which hydrocarbons become miscible with water.

Owing to the wide use of heavy-oil heating by steam injection [1, 2], Canada became one of the world's top ten hydrocarbon-producing countries. Heavy-oil recovery by steam injection requires two to five tonnes of steam per tonne of recovered oil [1-4]. After 20% of the heavy oil in place has been recovered from a reservoir, most of the steam energy is wasted to heating the water contained therein, and the profitability of production decreases by a factor of at least 2.

To improve crude-oil recovery, reservoir pressure is maintained by water injection into the reservoir. The adverse mobility ratio between oil and water results in early water breakthrough and low displacement efficiency. The average water content in reservoirs across Russia is 50%, and the average water cut from producing wells in Russia is 84%. When the percentage of oil in the produced fluid drops to (5–10) %, production is usually terminated. In such cases, the amount of remaining oil in place is comparable to the total amount of oil recovered. High reservoir water content is an increasingly serious problem in the modern petroleum industry [5]. The use of water injection to maintain reservoir pressure and of steam injection to produce high-viscosity oil only exacerbates the problem.

A.2 BINARY MIXTURE TECHNOLOGY FOR ENHANCING OIL RECOVERY AS AN ALTERNATIVE TO STEAM AND WATER INJECTION

The increase in reservoir water content is negligible when using a technology based on binary mixture reaction, which can now compete with advanced petroleum technologies.

Reservoir heating by binary mixture (BM) reaction is achieved by injecting solutions of two chemicals, ammonium nitrate and a nitrate decomposition initiator, into a well through separate passages [3]. The hot, high-pressure gas released at the bottom hole enters the reservoir. The main drawback of the early BM technique was a lack of control over down hole decomposition of ammonium nitrate, which would pose an explosion hazard when used in large amounts. For this reason, the amounts of ammonium nitrate injected in previous treatments have been restricted to one tonne per well. As ammonium nitrate decomposes, heat and gas are produced by the exothermic reaction

$$NH_4NO_3 \rightarrow N_2 + 2H_2O + 0.5O_2 + Q_1. \tag{1}$$

The products include oxygen, and its presence in a wellbore is permitted by Russian Technical Supervision Service safety regulations only if both temperature and pressure in the reaction zone are kept under control. The BM technique was improved in 2010/2011. We developed a reaction-control system to ensure safe injection of large amounts of NH_4NO_3 into reservoirs. The Emanuel Institute of Biochemical Physics of the RAS obtained permission for the use of unrestricted amounts of ammonium nitrate (Resolution no. 25-ID-19542–2010 of the Russian Technical Supervision Service).

A.3 EXPERIMENTAL REFINEMENT OF BINARY MIXTURE TECHNOLOGY: CONTROL AND OPTIMIZATION OF DOWNHOLE REACTION

In October/November 2011, an optimized down hole BM reaction was tested in wells 1242 and 3003 in the Usinsk oil field (LUKOIL–Komi

LLC). In particular, 15 tonnes of chemicals were injected into well 3003. Safe reaction progress at T = 200–300°C was achieved for the first time, while the recovery efficiency approached 0.8. The table below shows data on incremental oil recovery from both wells acquired by the end of March 2012.

Well no.	Pump model	Month	Production restart	Baseline daily production, t/d	Number of days	Average daily production, t/d	Incremental oil, tonnes
		Nov 2011	Nov 9, 2011	0.0	22	5.818	127.996
		Dec 2011			30.83	5.5	169.565
1242	EVNT-25	Jan 2012			31	4.63	143.0
		Feb 2012			29	4.94	143.26
		Mar 2012			31	3.98	123.38
		Total			143.83		**707.731**
		Jan 2012	Jan 4, 2012	1.93	28	10.6	242.9
3003	EVNT-25	Feb 2012			23	10.6	199.6
		Mar 2012			30.75	9.98	247.6
		Total			81.75		690.1

The total average daily production of wells 1242 and 3003 increased from 1.93 to 16.4 tonnes. The productivity index increased more than six-fold. During the 11 months after the treatment, wells 1242 and 3003 produced approximately 2500 tonnes of incremental oil. In December 2012, the total average daily production of wells 1242 and 3003 was about 8 tonnes.

A.4 THEORETICAL PROSPECTS OF BINARY MIXTURE TECHNOLOGY: HIGH-TEMPERATURE REACTION MAKES HYDROCARBONS MISCIBLE WITH WATER

It is well known from theory that heating water above Tc = 374°C at a pressure higher than Pc = 22 MPa causes a transition to a supercritical state, and the number of hydrogen bonds per water molecule decreases [4]. A significant amount of lighter hydrocarbon fractions also become supercritical and miscible with water, and the viscosity of the resulting water–oil mixture becomes a factor of tens or hundreds lower compared to that at average reservoir temperature (50–90°C) [4, 6]. This property of supercritical fluids can be used in a new enhanced oil recovery technique for bringing large amounts of water–oil mixture to the surface. It was proposed in [4] to use water in a supercritical state at Tc >374°C and Pc >22 MPa to make it miscible with oil.

It was shown in Refs. [3, 6, 7] that the reservoir stimulated by BM reaction must be partially saturated with gas produced by ammonium nitrate decomposition. The ensuing decrease in viscosity enables the reservoir fluid to flow toward the wellbore under a pressure gradient and up the wellbore to the surface under a gas-lift effect. An estimated 18.5 GJ of heat and at least 20 tonnes of gas were produced by Eq. (1) in wells 1242 and 3003 in the Usinsk oil field. The pumping power due to the gas-lift effect created during two hours after the injection of chemicals into well no. 3003 reached a maximum of 11.2 kW.

The cost of the chemicals consumed per tonne of incremental oil recovered from wells 1242 and 3003 is lower than 160 rubles (3.2$) [6]. The energy costs (electricity, oil, coal, natural gas) of generating 2–5 tones of steam (required to recover a tonne of bitumen) are several times higher.

A.5 THEORETICAL CALCULATIONS: A POSSIBLE REGIME OF BINARY MIXTURE REACTION AND EXTRACTION OF GAS-SATURATED OIL FROM A 250 Ч 250 M² RESERVOIR ZONE

Consider the following example of treatment using the heat and gas produced by ammonium nitrate decomposition. Suppose that the reservoir zone to be heated has a horizontal area of 250×250 m^2 and a height of 10 m with rock density of 2.5 t/m3, porosity of 20%, and rock specific heat of 1 MJ/t·K. The reservoir zone contains approximately 1.25×105 tonnes of oil whose specific heat is 2 MJ/t·K. The time required for the heated zone to cool down is at least 6 years as estimated by the method described in [3]. The projected time for recovering most of the heated oil is 18 months.

If reservoir rock temperature in the zone is to be raised by 200 K, then 3.125×105 GJ of heat must be added. Since approximately 5 GJ of heat is generated per tonne of ammonium nitrate decomposed, the required amount of nitrate is 6.25×104 tonnes. The total mass of the gas produced by decomposition (nitrogen, oxygen, and water vapor) is equal to the mass of ammonium nitrate consumed. Therefore, 0.5 t of gas must be released to heat one tonne of reservoir oil. This is sufficient to expect a rapid increase in well production rate even at an early stage of ammonium-nitrate injection into the reservoir.

Oil is expected to flow easily from the periphery through each cubic meter of the treated reservoir zone, displacing gas-saturated oil toward the wellbore. As a result, the reservoir is expected to yield approximately 1.25×105 tonnes of oil, while its temperature is expected to decrease by no more than 50 K [3].

Assuming that ten injection and production wells will have to be drilled, we estimate the profitability of production as follows.

Since wellhead prices in Russian oil fields vary between $200 and $300 per tonne, the 1.25×105 tonnes of oil projected to be recovered will make $25.0–$37.5M.

Lifting costs include:
- $6.25–9.375M for 6.25×104 tonnes of ammonium nitrate at $100–150 per tonne*.
- $1.75–7.0 M for drilling 10 wells to depths of no more than 500 m;
- $0.83M for operating costs.

These expenses add up to $9–18M. The gross profit estimated as the difference between sales and expenses should be at least $7M.

A.6 IMPORTANCE OF HETEROGENEOUS CATALYTIC REACTIONS ON RESERVOIR ROCK SURFACES

Our experience in treating terrigenous and carbonate reservoirs in 22 wells shows that the heating rate due to a thermochemical process depends both on the relative amounts of one tonne of reservoir oil. This is sufficient to expect a rapid increase in well production rate even at an early stage of ammonium-nitrate injection into the reservoir.

Oil is expected to flow easily from the periphery through each cubic meter of the treated reservoir zone, displacing gas-saturated oil toward the wellbore. As a result, the reservoir is expected to yield approximately 1.25×10^5 tonnes of oil, while its temperature is expected to decrease by no more than 50 K [3].

Assuming that ten injection and production wells will have to be drilled, we estimate the profitability of production as follows.

Since wellhead prices in Russian oil fields vary between $200 and $300 per tonne, the 1.25×10^5 tonnes of oil projected to be recovered will make $25.0–$37.5M.

Lifting costs include:
- $6.25–9.375M for 6.25×10^4 tonnes of ammonium nitrate at $100–150 per tonne*;
- $1.75–7.0 M for drilling 10 wells to depths of no more than 500 m;
- $0.83M for operating costs.

These expenses add up to $9–18M. The gross profit estimated as the difference between sales and expenses should be at least $7M.

A.7 IMPORTANCE OF HETEROGENEOUS CATALYTIC REACTIONS ON RESERVOIR ROCK SURFACES

Our experience in treating terrigenous and carbonate reservoirs in 22 wells shows that the heating rate due to a thermochemical process depends both on the relative amounts of chemicals utilized and the reservoir rock. We attribute this observation to heterogeneous catalysis on reservoir rock surfaces. This conclusion is consistent with the results of studies of catalytic down hole upgrading (e.g., see review in Ref. [8]).

The magnitude of thermochemical gas lift effect varies when the masses of chemicals injected into the heated reservoir are held approximately constant [4, 23]. To make the thermochemical gas lift effect instrumental in bringing reservoir fluid to the surface, the process should be made repeatable and controllable by injecting sufficiently strong catalysts into the reservoir. The catalytic activities of various rocks can be studied by using the technique developed in [9].

A.8 CONCLUSION

The estimates above suggest that profitable enhanced oil recovery is possible under a pressure gradient without adding water to the reservoir. The expected benefits include an increase in recovery factor and a reduction in operating costs due to shorter reservoir life.

Recovering approximately 2500 tonnes of incremental oil from wells 1242 and 3003 required a total of 10 tonnes of water contained in the injected BM solutions and produced by chemical reaction, whereas 5000 to 12,000 tonnes of steam must be injected to recover a similar quantity of incremental oil. Based on the results reported in [1–8] and obtained in tests performed in the Usinsk oil field, it can be stated that

- The injection of BM solutions into wells 1242 and 3003 took ten hours to complete, whereas the injection of several thousands of steam into a reservoir would last about a month;
- Steam injection is applicable only if the appropriate infrastructure is available, whereas BM injection can be performed by using readily deployable and inexpensive production equipment;
- Tests of a refined BM technique in the Usinsk oil field have shown that it can become an economical alternative to less efficient steam- or water-injection technologies;
- While the potential for further advancement of technologies that increase water cut is nearly exhausted, the prospects for developing the BM technology are just emerging.

In particular, a large-scale experiment is proposed to test the hypothesis of feasibility of replacing steam- and water-based technologies with BM technology. If successful, the proposed technology can improve the energy efficiency of oil production worldwide because increasing water content is a ubiquitous problem in reservoir engineering.

REFERENCES

1. Butler, R. M.; Stephens, D. J. J. *Can. Petrol. Technol.* **1981**, *20*, 90–96.
2. Butler, R. M.; Thermal Recovery of Oil and Bitumen; *Prentice-Hall: Englewood Cliffs*, **1991**.
3. Aleksandrov, E. N.; Kuznetsov, N. M.; Karotazhnik **2007**, *4*, 113–127 (in Russian).
4. Merzhanov, A. G.; Lunin, V. V.; Alexandrov, E. N.; Petrov, A. L.; Lidzhi-Goryaev, V. Yu. Nauka Tekhnol. Prom. **2010**, *2*, 1–6 (in Russian).
5. Griguletskii, V. G., *Tekhnol. TEK* **2007**. 10 (in Russian).
6. Aleksandrov, E. N.; Varfolomeev, S.D.; Lidzhi-Goryaev V. Yu.; Petrov, A.L. Tochka Opory **2012**, *15 (159)*, 4–5 (in Russian).
7. Aleksandrov, E. N.; Lemenovskii, D. A.; Koller, Z., Gas Evolving Oil Viscosity Diminishing Compositions for Stimulating the Productive Layer of an Oil Reservoir, *Patent WO 2010/043239*, April. 22, **2010**.
8. Kantzas, A.; Larter, S.; Araghi, B. M.; Pereira Almao, P. US 2010/0212893 A1 (patent application). August. 26, **2010**.
9. Aleksandrov, E. N.; Kozlov, S. N.; Kuznetsov, N. M.; Markevich, E. A.; Chastukhin, D. S. Comb. Expl. Shock Waves 2013, 49, 2–11.

CHAPTER 4

ANTIOXIDATIVE EFFICIENCY TERPENPHENOL AT THERMO-OXIDATIVE DEGRADATION POLYVINYL CHLORIDE

R. M. AKHMETKHANOV, S. V. KOLESOV, I. T. GABITOV, I. YU. HUKICHEVA, A. V. KUCHIN, and G. E. ZAIKOV

CONTENTS

4.1 INTRODUCTION

In the production of polymeric materials based on PVC to enhance their oxidative stability is used organic compounds with antioxidant function. A characteristic feature of this group is the ability of stabilizing to prevent or greatly inhibit the catalytic action of oxygen in energetic impacts of on the polymer [1].

The most effective antioxidants that increase the antioxidant stability of PVC and materials based on it, are phenol derivatives [2]. At present, a great scientific and practical interest of this class of compounds are ter-penphenols, particularly 4-methyl-2,6-izobornilphenol [3]. Known way to improve the thermal and photochemical stability of polypropylene with 4-methyl-2,6-izobornilphenol [4]. The use terpenphenols as PVC stabilizers unknown.

The aim of this work was to study the effects of certain terpenphenols on the process thermooxidative dehydrochlorination hard and plasticized PVC, the study of antioxidant activity the terpenphenols in autocatalytic oxidation of the plasticizers.

4.2 EXPERIMENTALS

Studied terpenphenols 4-methyl-2,6-izobornylphenol (struct 1), 6-methyl-2-izobornylphenol (struct 2), 4-methyl-2-izobornylphenol (struct 3).

Struct 1. 4-methyl-2,6-izobornylphenol.

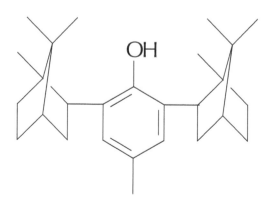

Struct 2. 6-methyl-2-izobornylphenol.

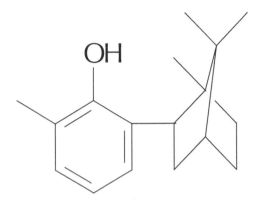

Struct 3. 4-methyl-2-izobornylphenol.

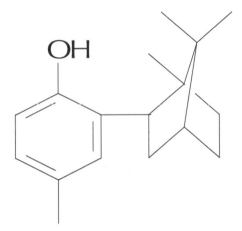

The 4-methyl-2,6-diizobornylphenol in rigid PVC in oxidative degradation leads to a marked reduction in the speed of dehydrochlorination of
the polymer (Fig. 1). The maximum decrease in the rate of elimination of
HCl is observed when the content terpenphenol 1.5–2.5 mmol/mol PVC.
Exceeding this value causes an increase in the speed of degradation of
the polymer. Decrease in the rate of dehydrochlorination thermooxidative
polymer in the presence terpenphenol observed almost to values corresponding to the value of the rate of thermal elimination of HCl from PVC
in an inert atmosphere, which is typical for stabilizers, antioxidants.

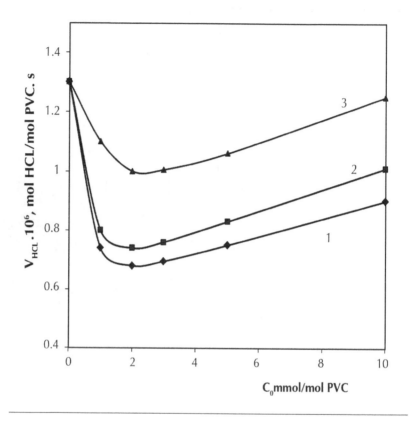

FIGURE 1   Dependence of the rate of thermal-oxidative dehydrochlorination of PVC content of antioxidants diphenylolpropane (1), 4-methyl-2 ,6-diizobornylphenol (2); ionol (3), (175°C, O_2, 3.5 l/h).

The stabilizing efficiency of 4-methyl-2,6-diizobornylphenol in terms of reducing the rate of dehydrochlorination of PVC thermooxidative al-most as good as the efficiency of industrial antioxidant - diphenylolpro-pane (2,2-bis(4-oxiphenyl)-propane) and significantly superior to ionol (4-methyl-2,6-ditertbutilphynol) (Fig. 1).

In the thermo-oxidative degradation of PVC, the introduction of the terpenphenols in hard polymer significantly reduces its rate of dehydro-chlorination (Fig. 2).

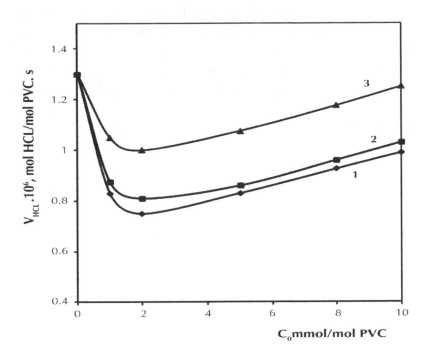

FIGURE 2 Dependence of the rate of thermal-oxidative dehydrochlorination of PVC content of antioxidants 6-methyl-2-izobornylphenol (1), 4-methyl-2-izobornylphenol (2); ionol (3), (175°C, O_2, 3.5 l/h).

The maximum decrease in the rate of elimination HCl is observed when the content terpenphenol 2 mmol/mol PVC. Exceeding this concentration increases the rate of degradation of the polymer.

As with the other phenolic antioxidants, reducing the rate of dehydrochlorination of the polymer in the presence of terpenphenols observed almost to values corresponding to the value of the rate of thermal elimination of HCl from PVC. As seen from (Fig. 2), the efficiency of stabilization 6-methyl-2-izobornylphenol and 4-methyl-2-izobornylphenol almost equal efficiency previously studied 4-methyl-2,6-diizobornylphenol, and significantly higher than the efficiency of industrial antioxidant – ionol.

The problem of stabilization of plasticized PVC is largely linked to the prevention of oxidative decomposition of plasticizers, in the presence of oxygen, in particular ester plasticizers, readily undergo free radical oxidation, activating the process of elimination of HCl from the polymer [5].

Thermooxidative dehydrochlorination process of PVC, plasticized dioctyl phthalate accompanied by autocatalysis (Fig. 3). The 4-methyl-2,6-diizobornylphenol in plasticized polymer leads to a sharp decrease in speed of thermal-oxidative dehydrochlorination of the polymer, and translation the process from the autocatalytic regime in to stationary.

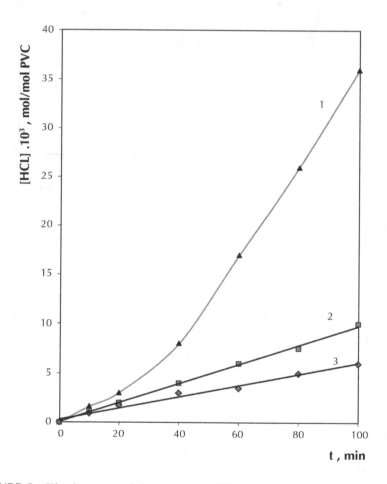

FIGURE 3 Kinetic curves of the process of dehydrochlorination of PVC plasticized with dioctyl phthalate (40 parts by weight/100 parts PVC) in the presence of 4-methyl-2 ,6-diizobornilphenol (the content of additives 1–0; 2–1; 3–2 mmol/mol PVC) (175°C, O_2, 3.5 l/h).

The maximum decrease in the rate of elimination of HCl from the polymer containing 40 parts by weight/100 parts PVC dioctylphthalate, as in the case not plasticized PVC content terpenphenols observed at 1.5–2.5 mmol/mol PVC. With more content terpenphenol is accelerated degradation of the polymer (Figs. 4–6).

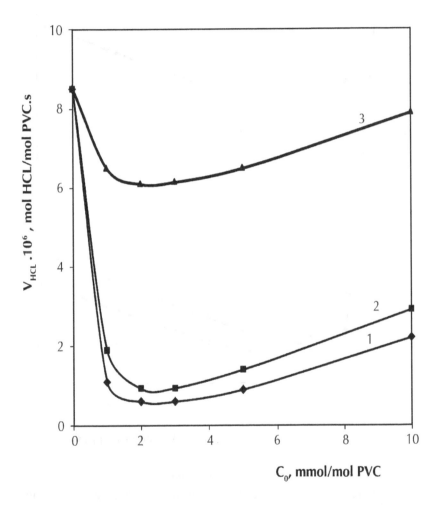

FIGURE 4 Dependence of the rate thermooxidative dehydrochlorination of PVC, plasticized dioctyl phthalate (40 parts by weight/100 parts PVC) on the content of antioxidants diphenylolpropane (1), 4-methyl-2,6-diizobornylphenol (2); ionol (3), (150°C, O_2, 3.5 l/h).

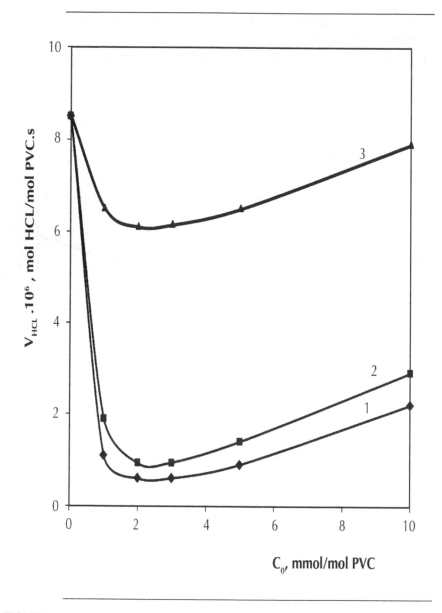

FIGURE 5 Dependence of the rate thermooxidative dehydrochlorination of PVC, plasticized dioctyl phthalate (40 parts by weight/100 parts PVC) on the content of antioxidants diphenylolpropane (1), 4-methyl-2, 6-diizobornylphenol (2); ionol (3), (175C, O_2, 3.5 l/h).

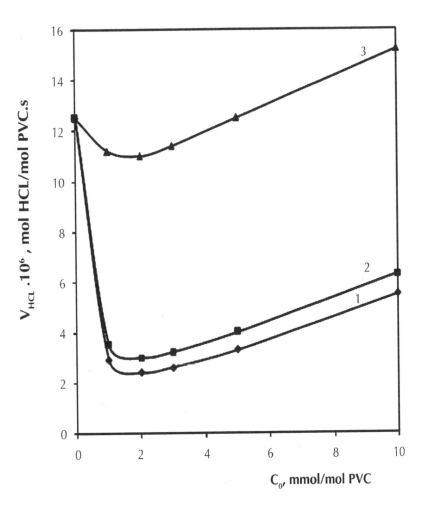

FIGURE 6 Dependence of the rate thermooxidative dehydrochlorination of PVC, plasticized dioctyl phthalate (40 parts by weight/100 parts PVC) on the content of antioxidants diphenylolpropane (1), 4-methyl-2, 6-diizobornylphenol (2); ionol (3), (190°C, O_2, 3.5 l/h).

A similar pattern is observed in oxidative degradation of PVC plasticized with dioctyl sebatsinat (40 parts by weight/100 parts PVC) degradation at different temperatures (Figs. 7–9).

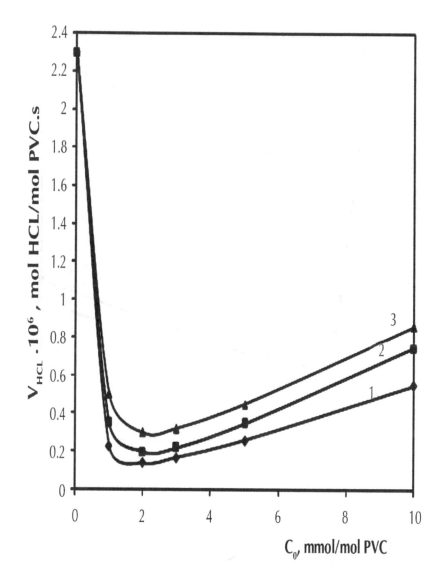

FIGURE 7 Dependence of the rate thermooxidative dehydrochlorination of PVC, plasticized dioctyl sebatsinat (40 parts by weight/100 parts PVC) on the content of antioxidants diphenylolpropane (1), 4-methyl-2,6-diizobornylphenol (2); ionol (3), (150°C, O_2, 3.5 l/h).

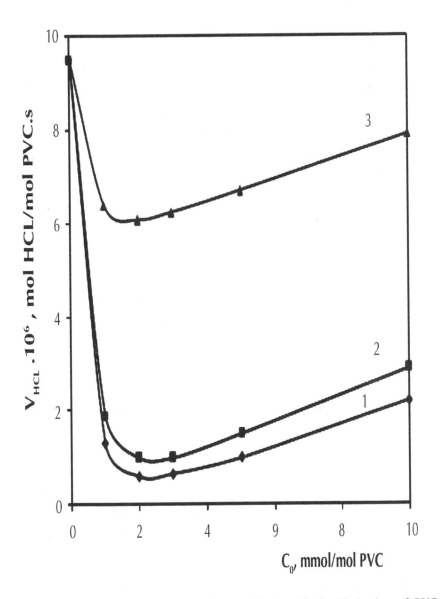

FIGURE 8 Dependence of the rate thermooxidative dehydrochlorination of PVC, plasticized dioctyl sebatsinat (40 parts by weight/100 parts PVC) on the content of antioxidants diphenylolpropane (1), 4-methyl-2,6-diizobornylphenol (2); ionol (3), (175C, O_2, 3.5 l/h).

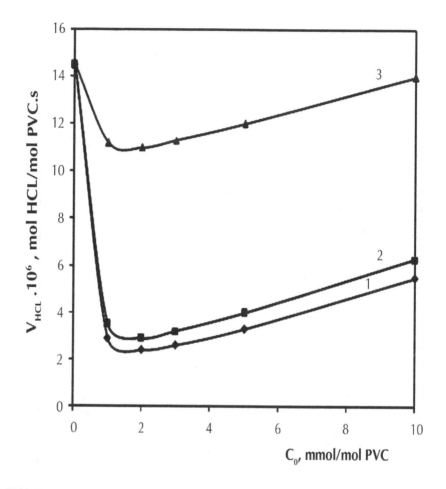

FIGURE 9 Dependence of the rate thermooxidative dehydrochlorination of PVC, plasticized dioctyl sebatsinat (40 parts by weight/100 parts PVC) on the content of antioxidants diphenylolpropane (1), 4-methyl-2,6-diizobornylphenol (2); ionol (3), (190°C, O_2, 3.5 l/h).

Speed reduction thermooxidative degradation of the polymer containing 4-methyl-2,6-diizobornylphenol observed values corresponding to oxidative degradation rate of unplasticized polymer. Obviously, terpenphenol plasticizer protects from oxidation, which in turn is due to solvation stabilization increases the thermal stability of PVC (known effect of "Echo stabilization") [6].

It should be noted that the maximum efficiency of anti-4-methyl-2,6-diizobornilfenola in terms of of relative velocity reduction dehydrochlorination ($\Delta V_{HCl}/V^0_{HCl}$) at different temperatures practically does not concede the stabilizing efficiency diphenylolpropane and significantly exceeds the high temperature degradation 175⁰C and 190⁰C efficiency of ionol (Table 1). The high efficiency of stabilizing terpenphenols compared to ionol degradation high temperature is probably related with low volatility supplements because terpenphenols has a higher melting temperature and a bulk chemical structure.

TABLE 1   The maximum the stabilizing efficiency of antioxidants in terms of of relative velocity reduction dehydrochlorination ($\Delta V_{HCl}/V^0_{HCl}$) at different temperatures degradation of PVC plasticized with dioctyl phthalate and dioctyl sebatsinat (40 parts by weight/100 parts PVC) *.

Antioxidants	$\Delta V_{HCl}/V^0_{HCl}$		
	150°C	175°C	190°C
4-methyl-2,6-diizobornylphenol	0.92/0.91	0.89/0.88	0.75/0.91
diphenylolpropane	0.93/0.94	0.91/0.90	0.80/0.81
ionol	0.88/0.89	0.17/0.29	0.11/0.12

* The numerator for the dioctyl phthalate, the denominator for dioctyl sebatsinat.

Considerably higher the stabilizing efficiency the studied terpenphenols exhibit at thermo-oxidative degradation of PVC, plasticized dioctyl phthalate (Fig. 10). The introduction of terpenphenols in PVC plastic compound causes a sharp reduction in the speed the dehydrochlorination of the polymer.

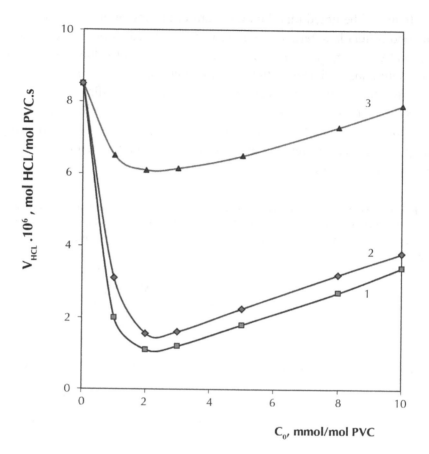

FIGURE 10 Dependence of the rate thermooxidative dehydrochlorination of PVC, plasticized (40 parts by weight/100 parts PVC) dioctyl phthalate in the content of 6-methyl-2-isobornylphenol (1), 4-methyl-2-isobornylphenol (2) and ionol (3), (175 C, O_2, 3.5 l/h).

The maximum decrease in the rate of decay of polyvinyl chloride, plasticized 40 parts by weight/100 parts PVC dioctyl phthalate, as in the case of destruction of the hard polymer is observed when the content terpenphenols 2 mmol/mol PVC. At high concentrations terpenphenols observed acceleration of dehydrochlorination of PVC.

Speed reduction thermooxidative dehydrochlorination of PVC in the presence terpenphenols observed up to the values of the speed to thermooxidative HCl elimination of rigid polymer. The stabilizing effect of terpenphenols because they protect the plasticizer from oxidation due to solvation stabilization, thereby increasing the thermal stability of the polymer.

The stabilizing efficiency 6-methyl-2-isobornylphenol, 4-methyl-2-isobornylphenol, as in the case of destruction of rigid PVC, almost equal efficiency 4-methyl-2,6-isobornylphenol, and significantly higher than the efficiency of ionol.

The high stabilizing the efficiency terpenphenols comparable to efficiency diphenylolpropane and exceeding of the ionol efficiency is also confirmed in terms the time thermal stability of PVC compounds. Additional introduction in plasticized PVC compounds containing metal-containing stabilizer – acceptor 4-methyl-2,6-diizobornilphenol, 4-methyl-2-isobornylphenol, and 6-methyl-2-isobornylphenol increases the rate of *"time thermal stability"* in 1.23–1.68 times (Table 2).

TABLE 2   The time value of thermal stability of PVC compounds.

Composition parts by weight/100 parts PVC	1	2	3	4	5	6	7	8	9	10	11
PVC	100	100	100	100	100	100	100	100	100	100	100
Dioctyl phthalate	40	40	40	40	40	40	40	40	40	40	40
Tribasic lead sulfate	1	1	1	1	1	1	1	1	1	1	1
2-Isobornyl-4-Methylphenol	-	0,05	0,1	-	-	-	-	-	-	-	-
4-Methyl-2,6-isobornyl phenol	-	-	-	0,05	0,1	-	-	-	-	-	-
6-methyl-2-isobornyl Phenol	-	-	-	-	-	0,05	0,1	-	-	-	-

TABLE 2   *(Continued)*

Diphenylol propane	-	-	-	-	-	-	-	0,05	0,1	-	-
Ionol	-	-	-	-	-	-	-	-	-	0,05	0,1
τ, min (175°C)	76	94	98	106	107	103	108	81	85	110	112

4.3   CONCLUSION

Thus, it is shown that 4-methyl-2,6-izobornilfenola, 6-methyl-2-isobornylphenol, 2-isobornyl-4-methylphenol in thermooxidative decomposition rigid and flexible PVC are effective antioxidants. degradation at temperatures above 150°C terpenphenols data on efficiency the stabilizing exceed ionol and almost as good as the efficiency of industrial antioxidant – difenilolpropan. The studied terpenphenols are of great interest for practical use in the production of polymeric materials based on PVC.

KEYWORDS

- 4-Methyl-2-izobornylphenol
- 4-Methyl-2,6-izobornylphenol
- 6-Methyl-2-izobornylphenol
- PVC
- Terpenphenols

REFERENCES

1. Minsker, K. S.; Fedoseeva, G. T. Degradation and stabilization of PVC. M.: Chem., **1979**, 272.
2. Shlyapnikov, Yu. A.; Kiryushkin, S. G.; Marin, A. P. Antioxidative stabilization of polymers. M.: Chem., **1986,** 256.
3. Chukicheva, I. Y.; Kuchin, A. V. Natural and synthetic terpenophenols. Russian. Chem. Well. **2004**, *48(3),* 2–37.
4. Mikhailov, N. V.; Tokareva, L. G.; Vlasov, A. V. The process of stabilization of poly-propylene and its products. AS USSR **1962**, *22(151), 769.*
5. Kulish E. I.; Kolesov S. V.; Minsker K. S. Effect of plasticizers on the thermal stabil-ity of the ester of polyvinyl chloride. Polymer Science. **2000**, *42(5),* 868–871.
6. Minsker, K. S.; Abdullin, M. I. The effect of "Echo-stabilization" in thermal PVC. *Reports of the Academy of Sciences of the USSR.* **1982**, *263(1),* 140–143.

CHAPTER 5

PHOTOINITIATED COPOLYMERIZATION OF BIFUNCTIONAL (METH)ACRYLATES— KINETICS AND MECHANISM

G. I. KHOVANETS', YU. G. MEDVEDEVSKIKH, I. YU. YEVCHUK, and G. E. ZAIKOV

CONTENTS

5.1   INTRODUCTION

On the basis of statistically significant experimental material onkinetics of photoinitiated copolymerization of bifunctional (meth)acrylates till high conversion depending on comonomer nature, system composition, photoinitiator concentration and intensity of UV radiation the kinetic model of three-dimensional copolymerization has been suggested according to the concept of microheterogeneity of polymerization system. In this chapter, we present the comparison of experimental and calculated kinetic curves and show their good coincidence. The analysis of calculated data and the effect of the obtained kinetic parameters on the general kinetics of the process has been discussed.

Theory of radical polymerization, particularly, of copolymerization, was developed over several decades and was created in its classic version by the end of the 1950s. A lot of theoretically predicted kinetic features have been confirmed experimentally. All these backgrounds are reflected in monographs [1–3].

At the same time the classic theory cannot generalize accumulated experimental results completely. Therefore the classic theory was supplemented, for example, with the notions about gel effect, microheterogeneity of polymerizing system, cage effect and diffusion control effect on polymerization features at high monomer conversion [4–13]. However, up to date, creation of the models completely describing the kinetics of (co) polymerization till high conversion of monomers by one position failed (kinetic features of (co)polymerization of monomers were studied only in the early stage of the process, while the main task is to study the process till high conversion).

In particular, above said may be applied also to photoinitiated radical (co)polymerization of mono- and polyfunctional monomers. Two diametrically-opposite concepts are used to describe the kinetics of photoinitiated (co)polymerization. The first one is the concept of diffusion-controlled reactions, based on the assumption about the diffusion control of elementary acts of macromolecular chain propagation and decay [14]. The second concept is the radical polymerization model, based on the notion of microheterogeneity of polymerizing system and different reaction zones. In each zone the process of polymerization occurs obeying its own laws [9].

In Refs. [15, 16], the kinetic models of radical (co)polymerization were obtained after the concept of polymerizing system microheterogeneity that considers the process of homopolymerization of polyfunctional monomers till high conversion in two reaction zones and of monofunctional monomers in three reaction zones.

We have experimentally studied the kinetics of copolymerization of mono- and polyfunctional monomers till high conversion for a number of mono- and di(meth)acrylates of different nature and functionality within the wide range of experimental conditions. On the basis of the obtained experimental data the kinetic model of linear and three-dimensional photoinitiated copolymerization of mono- and di(meth)acrylates till high conversion was proposed using the concept of microheterogeneity of polymerizing system and taking into account the process peculiarities in different reaction zones [17, 18]. The kinetic model of copolymerization at linear chain decay, which is necessary to analyze the kinetics of copolymerization till high conversion, was proposed as well, using the method of routes [19].

Comparison of the experimental kinetic curves and the curves, calculated using proposed model and analysis of the obtained results, presented in this chapter, show their good agreement.

5.2 KINETICS OF PHOTOINITIATED COPOLYMERIZATION OF DI(METH)ACRYLATES TILL HIGH CONVERSION

The kinetics of stationary (quazistationary after radicalconcentration in accordance with the Bodenstein—Semenov principle) photoinitiated copolymerization of bifunctional (meth)acrylates was studied by laser interferometry [11, 20]. The experimental kinetic curves were obtained for the systems: 1.6-hexanedioldiacrylate – triethylene glycol dimethacrylate (HDDA – TGM-3) at molar ratios of components HDDA: TGM-3 4:1, 2:1, 1:1, 1:2 and 1:4 in thin layers till high conversion depending on photoinitiator (2,2-dimethoxy-1,2-diphenylethane-1-on (IRGACURE

651)) concentration (1 and 2% mol.) and UV irradiation intensity (7, 17 and 48 W/m²) [18].

Preliminary purification of monomers was performed using technique, described in [15], namely, by mixing monomers with pre-activated Al_2O_3 powder and subsequent centrifugation. Purification was carried out in order to reduce the content of polymerization inhibitors. The absence of incubation period on the obtained kinetic curves testifies it.

The relative integral conversion P was estimated as the ratio between current contraction of photocomposition layer and its limit contraction (at $t \rightarrow \infty$). The experimental results are presented in the form of integral kinetic curves conversion time in Fig. 1.

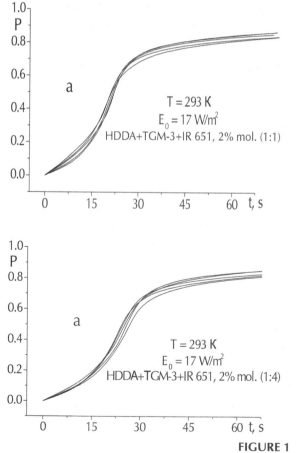

FIGURE 1   *(Continued)*

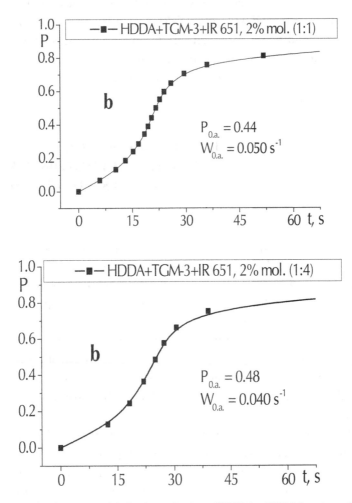

FIGURE 1 Kinetic curves of (co)polymerization of HDDA – TGM-3 systems (a) and the results of their averaging (b).

Fluctuation sensitivity of the process of di(meth)acrylates copolymerization must be noted. It affects the repeatability of the kinetic curves for the same conditions of the process (see Fig. 1). Therefore, for each condition of (co)polymerization several (usually 5–10) kinetic curves were got and then averaged as it is shown in Fig. 1.

Comparison of obtained averaged kinetic curves of copolymerization of di(meth)acrylates HDDA and TGM-3 at different intensities of UV irradiation depending on composition of copolymerizing system is shown in figures below.Kinetic regularities of photoinitiated copolymerization of the system HDDA – TGM-3 till high conversion were described and analyzed in [18].

5.3 COMPARISON OF THE EXPERIMENTAL AND CALCULATED DATA ON THE KINETICS OF PHOTOINITIATED COPOLYMERIZATION OF HDDA – TGM-3 SYSTEM TILL HIGH CONVERSION

On the basis of the experimental data general positions for derivation of general kinetic model of photoinitiated copolymerization of bifunctional (meth)acrylates till high conversion were formulated. They are as follows.

At (co)polymerization of bifunctional monomer a solid polymer phase, which begins to form at very low conversion (~ 1%), is a networked polymer; its solubility in a liquid monomer phase is small; monomer solubility in networked polymer is also small, so copolymerization process in polymeric phase may be neglected. A new reaction zone – an interphase layer on the boundary liquid monomer phase (MP) –solid polymer phase (PP) – is formed.

Thus, the process of polymerization of polyfunctional monomers is realized in two reaction zones: in volume of liquid monomer phase (MP), where dissolved polymer is absent practically, and in the interphase layer at the boundary of MP and solid polymer phase (PP), where monomer solubility is small. According to this model of two reaction zones the kinetic equations were derived [18]:

$$
W_i = -\frac{d[A_i]}{dt} = \frac{(k_{pii}k_{pij}[A_{vi}]^2 + k_{pij}k_{pji}[A_{vi}][A_{vj}])v_{in}^{1/2}}{\{k_{tii}(k_{pij}[A_{vi}])^2 + 2k_{tij}k_{pij}k_{pji}[A_{vi}][A_{vj}] + k_{tjj}(k_{pji}[A_{vj}])^2\}^{1/2}}(1-\phi_s)+
$$
$$
+\left(\frac{((k_{pii}k_{pij})_s[A_{vsi}]^2 + (k_{pij}k_{pji})_s[A_{vsi}][A_{vsj}])v_{in}}{(k_{tii}k_{pij})_s[A_{vsi}]^2 + (k_{tij}k_{pij}+k_{tji}k_{pij})_s[A_{vsi}][A_{vsj}] + (k_{tjj}k_{pji})_s[A_{vsj}]^2}\right)F_{vs}(1-\phi_s)\phi_s
$$

$$(1)$$

$$W_j = -\frac{d\left[A_j\right]}{dt} = \frac{(k_{pij}k_{pji}[A_{vj}]^2 + k_{pji}k_{pij}[A_{vj}][A_{vi}])v_{in}^{1/2}}{\{k_{tjj}(k_{pji}[A_{vj}])^2 + 2k_{tji}k_{pji}k_{pij}[A_{vj}][A_{vi}] + k_{tii}(k_{pij}[A_{vi}])^2\}^{1/2}}(1-\phi_s) +$$
$$+\left(\frac{\left(k_{pij}k_{pji}\right)_s[A_{vsj}]^2 + \left(k_{pji}k_{pij}\right)_s[A_{vsj}][A_{vsi}])v_{in}}{\left(k_{tjj}k_{pji}\right)_s[A_{vsj}]^2 + \left(k_{tji}k_{pji} + k_{tij}k_{pji}\right)_s[A_{vsj}][A_{vsi}] + \left(k_{tii}k_{pij}\right)_s[A_{vsi}]^2}\right)F_{vs}\left(1-\phi_s\right)\phi_s$$

(2)

$$\frac{d\phi_s}{dt} = \frac{W_i + W_j}{\left[A_i\right]^0 + \left[A_j\right]^0}$$

(3)

Here, k_{pij}, k_{tij} and k_{pijs}, k_{tijs} are rate constants of polymer chain propagation and decay in the volume of MP and in the interphase layer; $[A_{vi}]$, $[A_{vsi}]$ are concentrations of i-monomer in MP and in the interphase layer, respectively; V_{in} is an initiation rate; φ_s is current volume fraction of microgranules of PP; F_{vs} is a coefficient which takes into account fractal properties of a microgranule surface; $[A_i]^0$, $[A_j]^0$ are initial comonomer concentration.

The general conversion of monomers in copolymerization process was determined after the formula:

$$P = \frac{1}{2}\left(\frac{[A_i]^0 - [A_i]}{[A_i]^0} + \frac{[A_j]^0 - [A_j]}{[A_j]^0}\right)$$

(4)

The Eqs. (1) and (2) describe general rate of expense of each monomer (counting on the total system volume, with regard to that specific rates ω_i of the expense of i-monomer in each reaction zone are different and consist of two components:

(1) The first component of these equations is the known Mayo–Walling equation that describes the specific rate ω_{vi} of copolymerization in liquid MP by the classic kinetic model of polymerization at quadratic chain decay;

(2) The second component describes the specific rate ω_{vsi} of copolymerization on i-monomer in the volume of reaction zone of the interphase layer at linear chain decay.

Using the third equation, which was received from the balance equations and taking into account that monomers concentrations in PP are close to zero $[A_{si}] \approx 0$, $[A_{sj}] \approx 0$, the rate $d\varphi_s/dt$ of a solid polymer phase formation can be determined.

These models satisfactorily describe the experimental data in qualitative level, however, the numerical test was not done, so we have posed the problem – to solve so-called direct kinetic problem and to compare experimental and calculated kinetic curves.

The calculated kinetic curves were obtained at integration the presented system of differential Eqs. (1)–(4) by the Runge–Kutta 4th order method using mathematical program package Mathcad 2000 Professional. For convenience of calculation, we have converted molar-volumetric concentrations of components to relative current concentrations $X_i = [A_i]^0 / ([A_i]^0 + [A_j]^0)$, therefore, the initial monomer concentration is always equal to 1: $X_1^0 + X_2^0 = 1$; we have also combined the set of some constants of copolymerization into one value (as it is shown in Table 1) and accepted that concentrations of comonomers in MP are equal to concentrations of comonomers in an interphase layer $[A_{vsi}] = [A_{vi}]$ and $[A_{vsj}] = [A_{vj}]$. The obtained calculated data were compared with the experimental ones using computer program Origin 5.0.

TABLE 1　Accepted numerical values of constant kinetic parameters of photoinitiated copolymerization of HDDA and TGM-3.

$k_{p11} \cdot k_{p12}$	a_1	1
$k_{p12} \cdot k_{p21}$	a_2	0.01

TABLE 1   *(Continued)*

$k_{t11} \cdot k_{p12}^2$	a_3	2
$k_{t12} \cdot k_{p12} \cdot k_{p21}$	a_4	1
$k_{t22} \cdot k_{p21}^2$	a_5	1
$(k_{p11} \cdot k_{p12})_s$	a_6	20
$(k_{p12} \cdot k_{p21})_s$	a_7	0.8
$(k_{t11} \cdot k_{p12})_s$	a_8	0.002
$(k_{t12} \cdot k_{p12} + k_{t21} \cdot k_{p12})_s$	a_9	0.005
$(k_{t22} \cdot k_{p21})_s$	a_{10}	0.00025
$k_{p22} \cdot k_{p21}$	a_{11}	0.12
$k_{t21} \cdot k_{p21} \cdot k_{p12}$	a_{12}	1
$(k_{p22} \cdot k_{p21})_s$	a_{13}	0.7
$(k_{t21} \cdot k_{p21} + k_{t12} \cdot k_{p21})_s$	a_{14}	0.00025

In the values, for example, k_{p12} 1 –HDDA, 2 – TGM–3.

According to the obtained experimental data on the kinetics of stationary photoinitiated copolymerization of di(meth)acrylates HDDA and TGM-3 under different experimental conditions, the intensity of UV irradiation, which is equal to 7, 17 and 48 W/m² varies by approx. 2.4 and 6.8 times but this dependence is not reflected on the experimental kinetic curves (maximal copolymerization rate W_0 at the stage of acceleration (inflection point) at given intensities varies only by approx. 1.4 and 1.7 times). Taking into account this feature of copolymerization process, we have accepted the following numerical values of the initiation rate for calculations: $V_{in} = 5 \times 10^{-4}s^{-1}$, $7 \times 10^{-4}s^{-1}$ and $8.5 \times 10^{-4}s^{-1}$.

Comparison of the experimental integral kinetic curves and their differential anamorphoses of photoinitiated copolymerization of HDDA – TGM-3 system depending on its composition and intensity of UV radiation (a) and the corresponding calculated kinetic curves (b) is shown in Figs. 2–4.

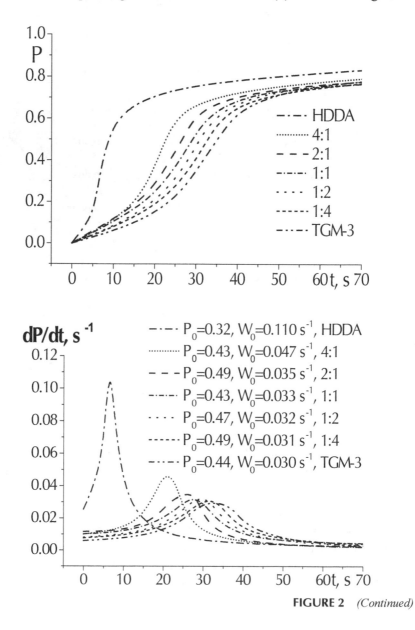

FIGURE 2   (Continued)

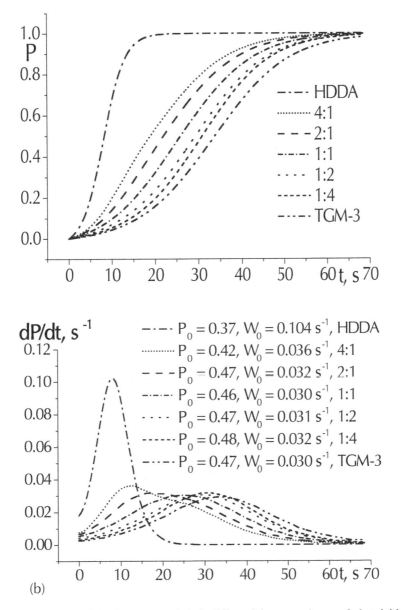

FIGURE 2 Integral kinetic curves and their differential anamorphoses of photoinitiated copolymerization of HDDA – TGM-3 system depending on its composition: [IR 651] = 2.0 % mol., T = 293 K, $E_0 = 7$ W/m² (a) and corresponding calculated kinetic curves: $V_{in} = 5 \times 10^{-4} s^{-1}$ (b).

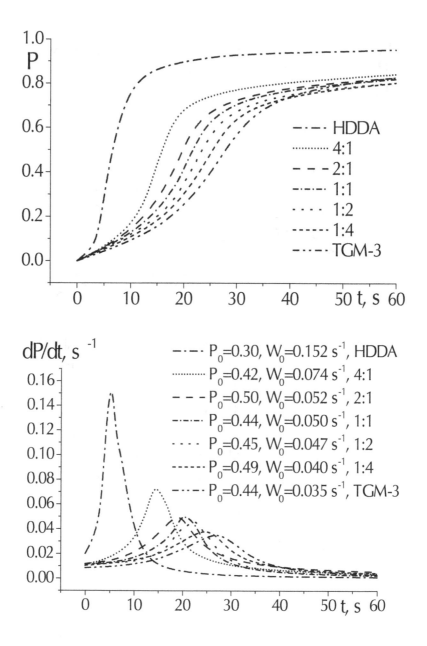

FIGURE 3   *(Continued)*

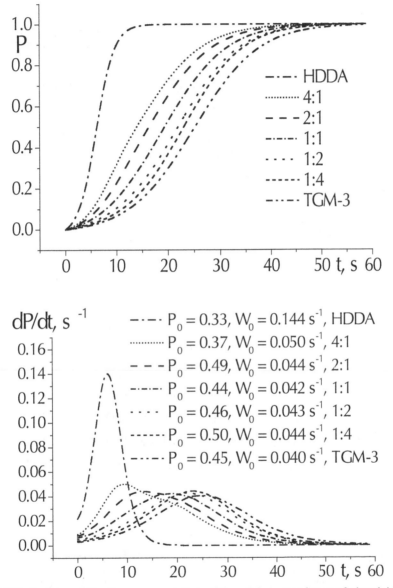

FIGURE 3 Integral kinetic curves and their differential anamorphoses of photoinitiated copolymerization of HDDA – TGM-3 system depending on its composition: [IR 651] = 2.0 % mol., T = 293 K, E_0 = 17 W/m² (a) and corresponding calculated kinetic curves: V_{in} = $7 \times 10^{-4} s^{-1}$ (b).

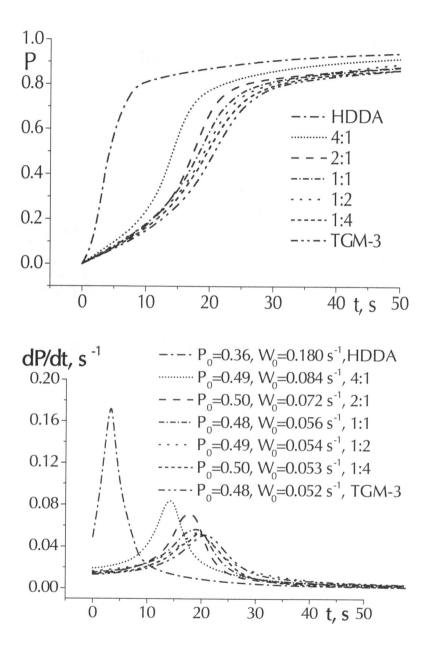

FIGURE 4　*(Continued)*

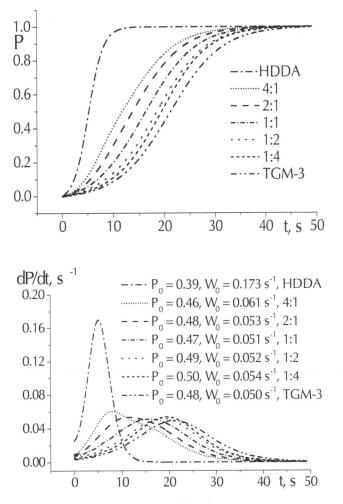

FIGURE 4 Integral kinetic curves and their differential anamorphoses of photoinitiated copolymerization of HDDA – TGM-3 system depending on its composition: [IR 651] = 2.0 % mol., T = 293 K, E_0 = 17 W/m² (a) and corresponding calculated kinetic curves: V_{in} = $8.5 \times 10^{-4} s^{-1}$ (b).

As one can see, integral experimental kinetic curves and corresponding calculated kinetic curves are qualitatively identical. They are of a typical S-shape and consists of only two parts: autoacceleration and autodeceleration. From the numerical data we can also see that accepted values of the initiation rate are in good correspondence with experimental dependence of copolymerization rate on UV irradiation intensity.

Presented experimental and calculated data clearly describe the features of copolymerization of HDDA : TGM-3 system, namely, that the process rate in it is strongly affected by the slower component (TGM-3). With the increase of TGM-3 fraction in the initial composition, conversion P_o immediately increases dramatically till limit values $P_o \cong 0,5$, that corresponds to the maximal rate W_o of copolymerization at the stage of autoacceleration, while W_o simultaneously decreases gradually.

At comparison of experimental and calculated kinetic curves one can observe the differences in the rates of achieving of maximum conversion at the final stage of autodeceleration. Obviously, at high conversion the process continues due to monomers frozen in polymeric matrix and radicals, that it not taken into account in the proposed model.

From the experimental data on kinetics of copolymerization of bifunctional monomers till high conversion we determined general conversion of copolymerization; we were unable to determine the expense of each monomer, while the calculated data allowed us to obtain numerical values of the rates of the expense of two components as well as the rate of solid polymer phase formation (Figs. 5 and 6), respectively.

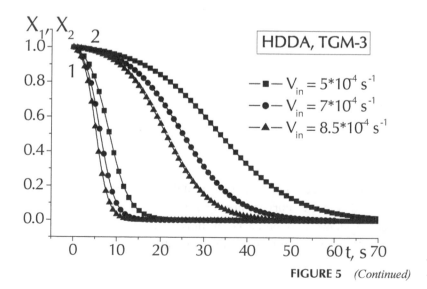

FIGURE 5   *(Continued)*

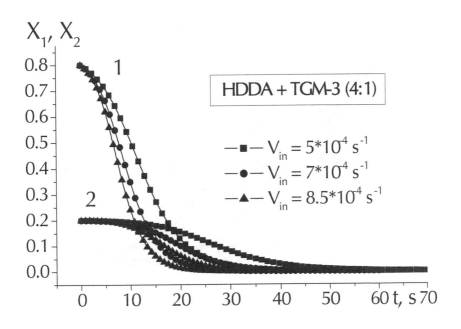

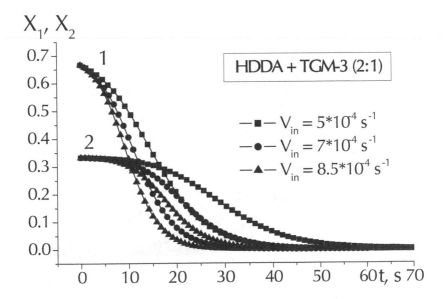

FIGURE 5 *(Continued)*

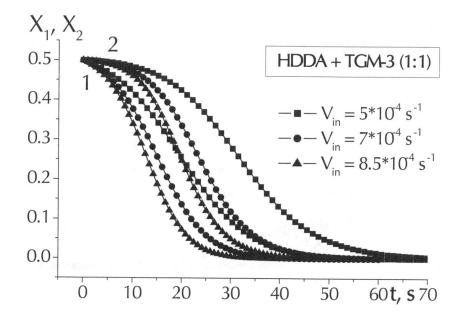

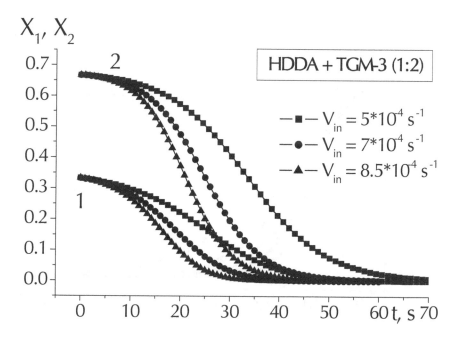

FIGURE 5 *(Continued)*

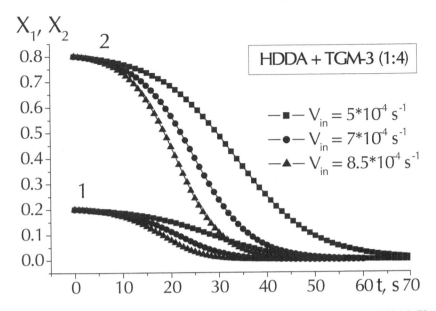

FIGURE 5   Calculated curves of monomer expense: 1 – HDDA (X_1) and 2 – TGM-3 (X_2) depending on the initial system composition.

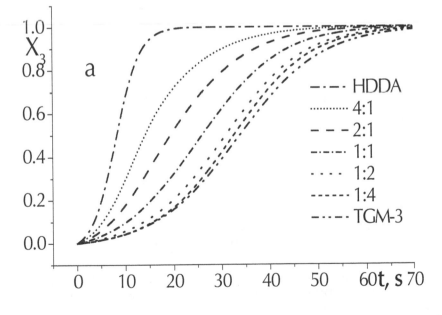

FIGURE 6   *(Continued)*

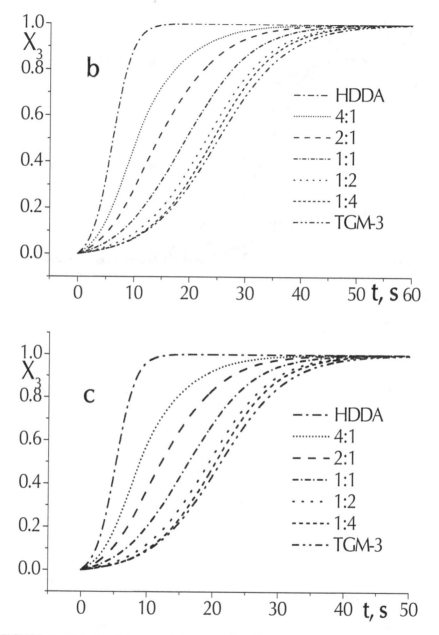

FIGURE 6 Calculated curves of the rate of solid polymer phase formation (X_3) at copolymerization of HDDA – TGM-3 system depending on its composition: $V_{in} = 5 \times 10^{-4} s^{-1}$ (a), $V_{in} = 7 \times 10^{-4} s^{-1}$ (b) and $V_{in} = 8.5 \times 10^{-4} s^{-1}$ (c).

As it is shown in Fig. 5, HDDA polymerizes much faster than TGM-3, but the regularities of copolymerization kinetics are determined by the slower component – TGM-3. At the beginning copolymer is enriched with the fast component and then with the slow one (see, f.e., Fig. 4 at component ratio 1:1 and $V_{in} = 5 \times 10^{-4}$ s^{-1}: at $t = 20$s HDDA has reacted by $\sim$ 50%, while TGM-3 only by $\sim$10%). Because of the difference in the rates of polymerization of the fast and slow components the copolymer composition changes dramatically during copolymerization as complex function of time.

In addition, kinetic models allow to change the ratio between the rate constant of polymer chain growth and decay both at homopolymerization and copolymerization and to observe the effect of this ratio on general kinetics of the process what is impossible during the experiment. All that allows anticipating and managing the process of copolymerization.

5.4 CONCLUSION

The analysis of the obtained experimental data on stationary kinetics of photoinitiated radical copolymerization of bifunctional (meth)acrylates allowed us to derive kinetic equations which describe the process of copolymerization till high conversion on the basis of the concept of microheterogeneity of polymerizing system. Integrating the system of differential equations by the Runge–Kutta 4th order method using program package Mathcad 2000 Professional we have calculated the numerical values of the rates of the expense of two components, the rate of solid polymer phase formation and general conversion.

Comparison of experimental and calculated kinetic curves showed their good correspondence. Analysis of the obtained numerical data shows that due to the difference in the rates of polymerization of the fast and slow components, copolymer composition changes during copolymerization and can be calculated by the model proposed.

KEYWORDS

- **Bifunctional monomers**
- **Comonomer nature**
- **Mono- and di(meth)acrylates**
- **Mono- and polyfunctional monomers**
- **Radical (co)polymerization**

REFERENCES

1. Riddle, E. H. Monomeric Acrylic Esters. New York: Reinhold, **1954**, 221.
2. Bemford, K.; Barb, J.; Jenkins, A.; Onion, P. Kinetics of radical polymerization of vinyl compounds. M.: "Inostr. lit.", **1961**, 348.
3. Bagdasaryan, H. S. Theory of radical polymerization. M.: "Nauka", **1966**, 300.
4. Ham, D. Copolymerization. M.: "Khimiya", **1971**, 616.
5. Oudian, J. Copolymerization. M.: "Mir", **1974, 654.**
6. Gladyshev, G. P.; Popov, V. A. Radical polymerization at high conversions. M.: "Nauka", **1974**, 244.
7. Spirin, Yu. L. Ractions of polymerization. K: "Naukova dumka", **1977**, 132.
8. Shur, A. M. Vysokomol. Soyed. M.: "Vys. shkola", **1981**, 656.
9. Berlin, A. A.; Koroliov, G. V.; Kefeli, T. Ya.; Sivergin, Yu. M. Acrylic oligomers and materials on their basis. M.: "Khimiya", **1983**, 232.
10. Ivanchev, S. S. Radical polymerization. L.: "Khimiya", **1985.** – 280.
11. Grishchenko, V. K.; Masliuk, A. F.; Gudzera, S. S. Liquid photopolymerizing composition. K.: "Naukova dumka", **1985**, 208.
12. Lavrov, N. A.; Sivtsov, E. V.; Nikolayev, A. F. Reaction medium and kinetics of polymerization processes. S.-P: "Sintez", **2001**, 94.
13. Medvedevskikh, Yu. G.; Kytsya, A. R.; Bazylyak, L. I., Turovsky, A. A.; Zaikov, G. E. Stationary and nonstationary kinetics of the photoinitiated polymerization. Utrecht, Boston: Brill Academic Publishers. **2004**, 313.
14. Kuchanov, S. I.; Povolotskaya, E. C. Proceedings of USSR Academy of Sciences, **1976**. *227(5),*. 1147–1150.
15. Zahladko, E.; Medvedevskikh Yu.; Turovski A.; Zaikov G. Intern. Journ. Polymer. Mater. **1998**, *39(3) & (4),*. 227–236.
16. Medvedevskikh, Yu.; Bratus A.; Hafiychuk H.; Zaichenko A.; Kytsya A.; Turovski A.; Zaikov G. J. Appl. Pol. Sci. **2002**, *86(14)*, 3556–3569.
17. Medvedevskikh, Yu.; Khovanets, G.; Yevchuk, I. Chemistry and Chemical Technology. **2009**, *3(1)*, 1–6.

18. Khovanets, G. I.; Medvedevskikh, Yu. G.; Yevchuk, I. Yu. Proceedings of Shevchenko Scientific Society. Chemistry and biochemistry. **2010,** 172–182.
19. Medvedevskikh, Yu. G.; Khovanets' G. I., Yevchuk, I. Yu.; Zaikov, G. E. Polymer Research Journal. **2008.** *1(3),.* 213–223.
20. Khovanets, G. I. Kinetic regularities of photoinitiated copolymerization of mono-and di(meth)acrylated till high conversion: Thesis for a degree of candidate of sciences on speciality 02.00.06. National Lviv Polytechnic University – Lviv, **2009,** 46–51.

CHAPTER 6

NANOFIBROUS ON THE BASE OF POLYHYDROXYBUTYRATE

A. A. OLKHOV, O. V. STAROVEROVA, YU. N. FILATOV,
G. M. KUZMICHEVA, A. L. IORDANSKY, G. E. ZAIKOV,
O. V. STOYANOV, and L. A. ZENITOVA

CONTENTS

6.1 INTRODUCTION

This chapter focuses on process characteristics of polymer solutions, such as viscosity and electrical conductivity, as well as the parameters of electrospinning using poly-3-hydroxybutyrate modified by titanium dioxide nanoparticles, which have been optimized. Both physical-mechanical characteristics and photooxidation stability of materials have been improved. The structure of materials has been examined by means of X-ray diffraction, differential scanning calorimetry (DSC), IR-spectroscopy, and physical-mechanical testing. The fibrous materials obtained can find a wide application in medicine and filtration techniques as scaffolds for cell growth, filters for body fluids and gas-air media, and sorbents.

Development of materials with revolutionary characteristics is closely connected with obtaining nanosized systems. Of greatest interest today are compositions derived from polymers and nanosized objects, which show a unique set of characteristics, have no counterparts, and drastically change present ideas about a polymer material.

Titanium dioxide nanoparticles are the most attractive because of the developed surface of titanium dioxide, the formation of surface hydroxyl groups with high reactivity resulted from reacting with electrolytes as crystallite sizes decrease down to 100Å and lower, and a high efficiency of oxidation of virtually any organic substance or many biological objects.

Many modern applications of TiO_2 are based on using its anatase modification, which shows the minimum surface energy and greater concentration of OH-groups on sample surface compared with other modifications. According to Dadachov's patents [1, 2], the nanosized η-modification of TiO_2 is considerably superior to anatase in the above-mentioned properties. The main characteristics of titanium dioxide modifications in use are sample composition, TiO_2 modification, nanoparticle size and crystallite size, specific surface area, pore size and pore volume.

Polyhydroxybutyrate (PHB) is the most common type of a new class of biodegradable termoplasts, namely polyoxyalkanoates. It demonstrates a high strength and the ability to biodegrade under natural environmental conditions, as well as a moderate hydrophilicity and nontoxicity (biodegrades to CO_2 and water) [3]. The PHB shows a wide range of useful performance characteristics [4]; it is superior to polyesters, which are the standard materials for implants, can find application in different branches of medicine, and is of great importance for cell engineering due to its bio-compatibility [5].

PHB is a unique sample of a moderate hydrophobic polymer being biocompatible and biodegradable at both high melting and crystallization temperatures. However, its strength and other characteristics, such as thermal stability, gas permeability, and both reduced solubility and fire resistance, are insufficient for its large-scale application.

The objective of the research was to prepare ultra-fine polymer composition fibers based on polyhydroxybutyrate and titanium dioxide and to determine the role played by nanosized titanium dioxide modifications in achieving special properties of the compositions.

6.2   EXPERIMENTAL

The nanosized η-TiO_2 and anatase (S12 and S30) were prepared by sulfate process from the two starting reagents, $(TiO)SO_4 \cdot xH_2SO_4 \cdot yH_2O$ (I) and $(TiO)SO_4 \cdot 2H_2O$ (II) correspondingly [6].

Samples of titanium dioxide and its compositions with polymers were analyzed by X-ray diffraction technique, using HZG-4 (Ni filter) and (plane graphite monochromator) diffractometers, CuK_α radiation, diffract-ed beam, in the range of 2θ 2–80°, rotating sample, stepwise mode (the impulse accumulation time is 10 s, by step of 0.02°). Experimental data array was processed with PROFILE FITTING V 4.0 software. Qualitative phase analysis of samples was carried out by using JCPDS PDF-2 database, ICSD structure data bank, and original papers.

Particle sizes (coherent scattering region) of TiO_2 samples were calculated by the Selyakov–Scherrer equation $L = \dfrac{k\lambda}{\beta \cdot \cos\theta_{hkl}}$, where $\beta = \sqrt{B^2 - b^2}$ is physical peak width for the phase under study (diffraction reflections were approximated by Gaussian function), B is integral peak width, b is instrument error correction ($b \sim 0.14°$ for α-Al_2O_3 as a reference), $k \sim 0.9$ is empirical coefficient, λ is wavelength. Calculations were based on a strongest reflection at $2\theta \sim 25°$. Standard deviation was $\pm 5\%$.

Starting PHB with molecular weight of 450 kDa was prepared through microbiological synthesis by BIOMER (Germany). Chloroform was used as solvent for preparing polymer solution. Both HCOOH and $[CH_3(CH_2)_3]_4N$ were used as special additives.

Electrostatic spinning of fibers based on PHB and titanium dioxide was carried out with original laboratory installation [7].

The dynamic viscosity of polymer solutions of various compositions was measured as a function of PHB concentration with Heppler and Brookfield viscosimeters. The electrical conductivity of polymer solutions was calculated by the equation $\lambda = a/R$, $Ohm^{-1}cm^{-1}$; the electrical resistance was measured with E7-15 instrument.

The fiber diameter distribution was studied by microscopy (optical microscope, Hitachi TM-1000 scanning electron microscope). Fiber orientation was studied by using birefringence and polarization IR-spectroscopy (SPECORD M 80 IR-spectrometer). Crystalline phase of polymer was studied by differential scanning calorimetry (DSC) (differential scanning calorimeter). The packing density of fibrous materials was calculated as a function of airflow resistance variation with a special manometric pressure unit [8].

Physical-mechanical characteristics of fibrous materials were determined with PM-3-1 tensile testing machine according to TU 25.061065-72. Kinetics of UV-aging was studied with Feutron 1001 environmental test chamber (Germany). Irradiation of samples was carried out with a 375 W high pressure Hg-lamp, at a distance of 30 cm.

For *in vitro* biodegradation studies, the materials were incubated in test tubes filled with 10 ml of 0.025 M phosphate buffer solution (pH = 7.4) at 70°C for 21 days. At regular time intervals the materials were removed from the buffer and rinsed with distilled water, then placed into incubator at 70°C for 3 hours, and finally weighed within 0.001 g.

6.3   RESULTS AND DISCUSSION

Diffraction patterns of the nanosized anatase and η-TiO_2 prepared are presented in Figs. 1a and 1b correspondingly. The sample S30 referring to anatase contains trace amounts of β-TiO_2 (JCPDS 46-1238) (Fig. 1a).

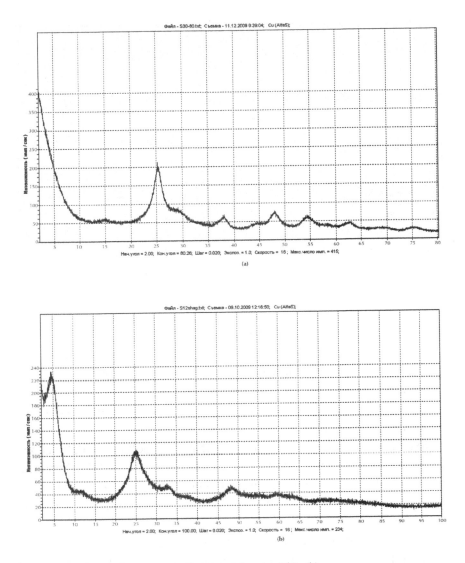

FIGURE 1   Diffraction patterns of anatase (a) and η-TiO$_2$ (b).

Analysis of the obtained size values of coherent scattering regions (*L*-values) of the η-TiO$_2$ and anatase samples showed that *L* = 50 (2) Å and *L* =100 (5) Å, i.e., crystallite sizes for the η-TiO$_2$ samples are substantially less.

Time dependencies of the dynamic viscosity of PHB solutions are shown in Fig. 2.

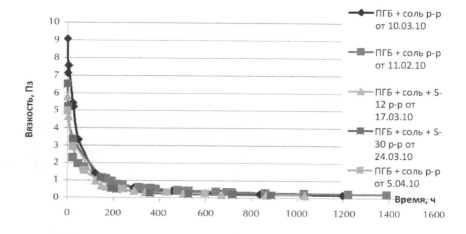

FIGURE 2   Dynamic viscosity as a function of PHB solution concentration.

Analysis of the dynamic viscosity of polymer solutions showed how the solution viscosity varies with time as the (HCOOH) – (S1) is added. Most probably, the decrease in viscosity results from the decrease in the molecular weight of polymer.

Examination of the electrical conductivity of polymer solutions and choosing solvent mixture allowed the $([CH_3(CH_2)_3]_4N)$ – (TBAI) concentration in solution to be decreased from 5 down to 1 g/l due to adding the S1. The increase in PHB concentration up to 7 wt. % in a new solvent mixture was gained.

The fiber diameter distribution was studied by using fibers from 5% PHB solution in chloroform/formic acid mixture and 7% PHB solution as well. It was found that 550–750 nm diameter fibers are produced from 5% solution, while 850–1250 nm diameter fibers are produced from 7% solution, i.e., the fiber diameter increases as the solution concentration rises (Fig. 3). The increase in the process velocity leads to the fiber diameter virtually unchanged.

(a)

(b)

(c)

FIGURE 3 *(Continued)*

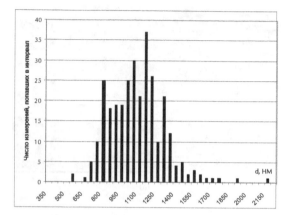

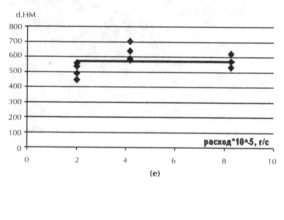

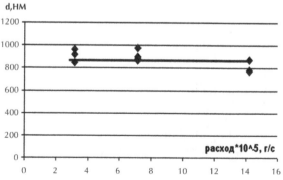

FIGURE 3 Microscopic images of fibrous material as a function of PHB solution concentration: (a) 5%, (b) 7%; fiber diameter distributions: (c) 5%, (d) 7%; the fiber diameter as a function of the process velocity (flow rate): (e) 5%, (f) 7%.

Examination of fibrous materials by electron microscopy showed that the fibrous material uncovered by the current-conducting layer decomposes in 1–2 min, when exposed by electron beam.

Examination of polymer orientation in fibers by using birefringence showed that elementary fibers in non-woven fabric are well oriented along fiber direction. However, it is impossible to determine the degree of orientation quantitatively because the fibers are packed randomly.

Analysis of fibrous materials by differential scanning calorimetry showed that at low scanning rates a small endothermic peak appears at 190/200°C, which indicates the presence of maximum straightened polymer chains (orientation takes place). Upon polymer remelting this peak disappears.

IR-examination showed that this method could be also applied only for qualitative estimation of orientation occurrence.

Measurement of the packing density of fibrous materials showed that the formulation used a day after preparing the solution has the greatest density of fiber packing. Probably, it is concerned with a great fiber diameter spread when smaller fibers are spread between bigger ones. The TiO_2-containing fibers are also characterized by the increased packing density.

The results of powder diffraction analysis of PHB powder (a) and PHB fibers obtained from 7% solution with modifying additives (b) and the nanosized TiO_2 modifications (c, d) are presented in Fig. 4.

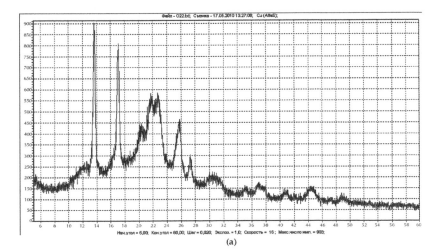

(a)

FIGURE 4 *(Continued)*

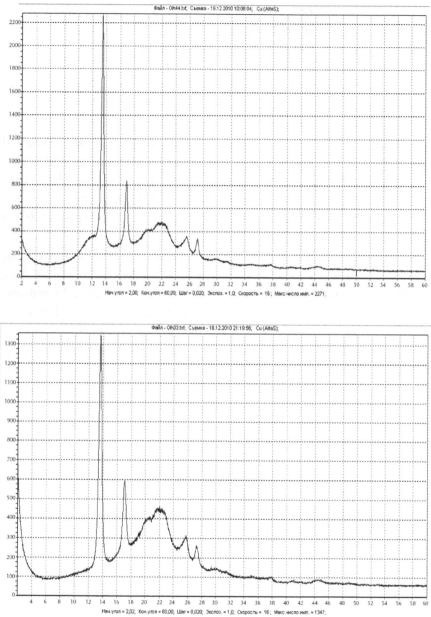

(c)

FIGURE 4　(Continued)

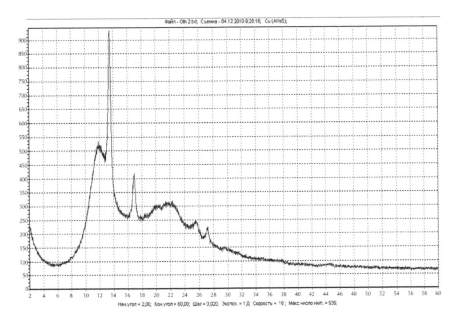

FIGURE 4 Diffraction patterns of: PHB (a), PHB-B (7% PHB+1% TBAI) (b), PHB-B +S 30 (c), PHB-B + S12 (d).

The results of physical-mechanical testing of fibrous materials of different formulations are presented in Table 1.

TABLE 1 Physical-mechanical characteristics of fibrous materials prepared from polymer compositions as a function of formulation.

Formulation	L, m	ε, %
PHB + TBAI	724.43	20.83
PHB + TBAI (after a day)	1001.78	12.48
PHB + TBAI + S-12	1430.45	53.37
PHB + TBAI + S-30	1235.91	62.02

where, L is breaking length, ε is breaking elongation.

Physical-mechanical tests showed that introducing nanosized TiO_2 into solution substantially alters the properties of the resulting fibrous material.

Non-woven fibrous material contains randomly packed fiber layers. Actually, its deformation can be considered as a "creeping", and more and more fibers while straightening during deformation contribute to the breaking stress growing up to its maximum. In other words, the two processes take place simultaneously, namely fiber straightening and deformation of the fibers straightened.

Obviously, as the TiO_2 content increases the fiber flexibility rises due to the decreased crystallinity. This results in a greater amount of fibers simultaneously contributing to the breaking stress and, consequently, the greater slope of the curve.

Moreover, the addition of both S-12 and S-30 is likely to cause the formation of a firm fiber bonding at crossing points, possibly due to hydrogen bonding. Figuratively speaking, a network is formed. At the beginning, this network takes all the deformation stress. Finally, network breaking occurs followed by straightening and "creeping" of fibers, which does not affect the stress growth.

There are different additives used in solutions. By purpose, all additives can be divided into process additives and production additives. The first ones are used to control both viscosity and electrical conductivity of spinning solutions as well as the velocity of fiber formation. The second ones are intended for obtaining fibrous products with desired properties.

Examination of the crystallinity of fibrous materials, films, and PHB powder showed that main crystallite modification of both PHB powder and fibers and films melt at 175–177°C, i.e., the morphology of crystals remains unchanged.

For fibers, the low-melting shoulder at 160–163°C appears which confirms either the presence of smaller crystallites or their imperfection. For the S-12 based formulation, the low-melting peak disappears which is the evidence of more uniform distribution of the additive within the material.

The broadened crystallization peak is observed for the S-12 based samples. The increased friction and decreased chain mobility lead to the obstructed crystallization and the decreased crystallization rate.

The narrowed crystallization peak is observed for the S-30 based samples. Crystallization proceeds only where no contact with this additive is. Hence, the S-12 and S-30 have different energies of intermolecular interaction of PHB chains and the additive surface.

The crystallinity was calculated by the following equation:

$$\alpha_{kp} = \frac{H_{n\pi}}{146},$$

where, $H_{n\pi}$ is melting heat calculated from melting peak area, J/g; 146 is melting heat of monocrystal, J [9].

The most loosely packed structure, i.e., the most imperfect, is observed for the formulation prepared a day after.

The S-12 based samples demonstrate excessive fiber bonding; the TiO_2 particles themselves obstruct crystallization. In the case of the S-12, the distribution is good.

The S-30 obstructs crystallization to a lesser extent; the fibers crystallize worse, and the chains are not extended.

Crystallization in fibers occurs upon orientation. The additive is not considered as a nucleating agent Table 2.

TABLE 2   Widths of crystallization peaks at the scanning rate of 20°C/min.

Material	Heating No.	Tm, °C	Crystallization peak width (at the onset), °C
PHB (starting)	1	174.9	18.61
	2	173.97	18.02
PHB fibers 1g/l TBAI	1	176.83	20.9
	2	169.27–159.74	20.93
PHB fibers 1g/l TBAI a day after	1	176.94	26.24
	2	171.98–163.14	22.8
PHB fibers 1g/l TBAI based on S-12	1	177.49	34.11
	2	170.33	27.38
PHB film from chloroform solution	1	177.21	18.27
	2	171.95–163.1	18.57

UV-aging tests showed that TiO_2-modified fibers demonstrate greater UV-aging resistance. Although the induction period for S-30 based samples is less than that for other formulations, the UV-degradation rate for the S-30 based sample is comparable to that for the S-12 based one.

For TiO_2-containing samples the increased thermal degradation heat after UV-aging is characteristic since the UV-treated TiO_2 acts as the initiator of both thermal and thermooxidation degradation due to OH-groups traveling to powder granule surface.

The onset temperature of both thermal and thermooxidation degradation decreases due to UV-degradation proceeded not only within amorphous phase but also within crystalline phase, as said above.

The TiO_2 acts as the initiator in UV-aging processes.

6.4 CONCLUSION

- Physical-mechanical characteristics of fibrous materials increase as the TiO_2 is introduced.
- The morphology of main PHB crystallites in powder and fibers is kept unchanged. However, the fibers show the low-melting shoulder (small and imperfect crystals). The TiO_2 obstructs crystallization.
- PHB fibers are characterized by strongly pronounced molecular anisotropy.
- S-12 based samples show the best thermal- and thermooxidation degradation stability as well as UV-aging resistance.
- The results obtained can be considered as the ground for designing new biocompatible materials, such as self-sterilizing packing material for medical tools or a support for cell growth.

KEYWORDS

- **Differential scanning calorimetry**
- **Electrical conductivity**
- **Poly(3-hydroxybutyrate)**
- **Polyhydroxybutyrate**
- **Titanium dioxide nanoparticles**
- **Viscosity**

REFERENCES

1. Dadachov, M. *U.S. Patent Application Publication.* US 2006/0171877.
2. Dadachov, M. *U.S. Patent Application Publication.* US 2006/0144793.
3. Fomin, V. A.; Guzeev, V. V.; *Plasticheskie Massy.* **2001**, *2.* 42–46.
4. Boskhomdgiev, A. P. Author's abstract of thesis for Ph.D. in Biology. Moscow, **2010**.
5. Bonartsev, A. P. et al. *J. of Balkan Tribological Association.* **2008**, *(14),* 359–395.
6. Kuzmicheva, G. M.; Savinkina, E. V.; Obolenskaya, L. N.; Belogorokhova, L. I.; Mavrin, B. N.; Chernobrovkin, M. G.; Belogorokhov, A. I. *Kristallografiya.* **2010**, *55(5),* 913–918.
7. Filatov, Yu. N. Electrospinning of fibrous materials (ES-process)/Moscow: Khimiya, **2001**, 231.
8. Barham, P. J.; Keller, A.; Otum, E. L.; Holms, P. A. *J. Master Sci.* **1984**, *19(27),* 81–279.

CHAPTER 7

DEGRADATION OF SOME FILMS UNDER UV-RADIATION

A. A. OLKHOV, A. L. IORDANSKY, G. E. ZAIKOV,
O. V. STOYANOV, L. A. ZENITOVA, and S. YU. SOFINAPETRONIS

CONTENTS

7.1   INTRODUCTION

The effect of poly(3-hydroxybutyrate) on expenditure and accumulation of chromophore groups, as well as the absorption of oxygen during photo-oxidation of vinyl alcohol with vinyl acetate copolymer was studied.

Photooxidation and photodestruction of polymeric materials are some of the most important factors determining the life of products from these materials and the ability to decompose under natural conditions at the end of exploitation.

In this regard, the natural polymer poly(3-hydroxybutyrate) (PHB), which has high biocompatibility and biodegradability, is of considerable interest. Relatively low physical and mechanical properties of this polymer are its well-known shortcomings. A large number of studies on the development and study of materials based on blends of PHB with other polymers [1, 2], mainly polyethylene [3, 4], is due to this fact. Additional interest in such mixtures is associated with the development of new materials for medical purposes, in which the use of PHB, as well as other moderately hydrophilic polymers [5], can regulate diffusion and sorption characteristics of such materials, in order to control the rate of drugs release in the body.

Selection of mixtures on the base of vinyl alcohol with vinyl acetate copolymer (VAVAC) and PHB as the objects of the study is due to several factors. Very limited compatibility of PHB with polymers of other classes, along with the high crystallinity of the PHB, leads to a sharp increase in light scattering of such mixtures, even with the use of optically transparent polymers selected as the second component. This greatly complicates the quantitative analysis of the photooxidation processes of the mixtures based on PHB, important for understanding the role of the main reactions that determine the stability of these materials in the natural environment. In contrast to many other systems in this regard mixture of VAVAC and PHB with the low ($\leq 20\%$) content of PHB has fairly good transparency, which makes them a promising object of research. The principal added incentive is the availability of data on the partial compatibility of phases in such mixtures [6], which allows counting on the significant effects of the mutual influence of the components. From a practical point of view, the interest in their research related to the prospective of their use as biocompatible materials for medical purposes, as well as packaging materials, suitable in particular for food products.

7.2   EXPERIMENTAL

The PHB produced by microbial synthesis by BIOMER® company (Germany) as a white fine powder (molecular weight of approx. 340,000, the melting point of 176°C, the degree of crystallinity of 69%), and industrial VAVAC grade 8/27 (Russia) with vinyl acetate content of 27% and molecular weight of 38,000 were used in the study.

From mechanical mixtures with a given weight ratio of the components films with the thickness of 60±5 or 200±10 micrometers were molded by the single screw extruder ARP-20 (Russia) with a screw diameter of 20 mm and a diameter to length ratio 25. Temperature of the extruder zones was varied from 150 to 190°C.

The samples were irradiated in the air by mercury lamp with high (DRSh-1000) or low pressure (DB-60), as well as in the machine for accelerated light stability tests "SUNTEST XLS+," in which the radiation is almost completely corresponds to the solar radiation in nature conditions.

The rate of oxygen consumption was determined by the manometric method on a special installation with thermostated quartz cell having a sensitivity of about 2×10^{-8} mole. Samples temperature during irradiation was kept constant with accuracy 0.1°C using a water thermostat. As an absorber of oxidation products solid KOH was used.

The absorption spectra of the original films, as well as their changes during irradiation were recorded using the spectrophotometers "Specord UV-Vis" and "MultiSpec-1501."

The irradiation of films from the mixtures of VAVAC and PHB as well as films from individual VAVAC, by the light with $\lambda=254$ nm of a mercury low pressure lamp DB-60 or by the light with $\lambda>290$ nm of a xenon lamp in the machine "SUNTEST XLS +", leads to a complex change in the absorption spectra in the near UV- and visible area: at the beginning of irradiation optical density sharply decreases and then slowly increases approaching with the prolonged irradiation to its stationary value.

7.3   RESULTS AND DISCUSSION

As shown in Fig. 1, the period of irradiation, during which the optical density reaches its initial value depends strongly on the composition of the samples and this dependence the smaller, the higher the content of PHB in the blend.

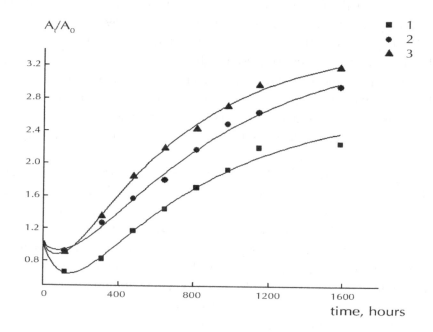

FIGURE 1   Changes in the relative optical density in the area of 357 nm during irradiation in air by the light of a mercury low pressure lamp DB-60 with $\lambda=254$ nm of the films with the thickness of 0.2 mm from VAVAC (1) and from the mixtures of VAVAC with 10 (2) or 20% PHB (3). Points – experimental data, curves – calculations by Eq. (1).

This effect is especially well noticeable when comparing samples from a mixture of VAVAC and PHB (20%) and individual VAVAC (Fig. 1, curves 3 and 1). Apparently, this is due to a higher rate of formation of chromophore groups in the presence of PHB. A higher value of a stationary optical density of the sample of the polymers mixture is also due to this reason.

X-ray diffraction data, and also the dependencies of the melting and glass temperatures from the composition, obtained in Ref. [6], show the compatibility of PHB and VAVAC. The interaction, however, is limited, which is manifested, in particular, in the presence of pronounced characteristic reflections of crystalline phase of PHB in the diffractograms of films containing ≥30% of PHB. Based on the dependencies analysis of physical and mechanical properties (tensile stress, elongation at break and modulus of elasticity), and also the coefficient of water diffusion from the ratio of the components it was found that when the content of PHB in the blend is up to 20% the continuous phase forms VAVAC partially modified by poly(3-hydroxybutyrate) [6]. PHB rather easily oxidizes under the short UV light [7], and acts as a high molecular photoinitiator in the mixture with VAVAC, which accelerates a photo transformation of VAVAC with the formation of colored products.

The results of the study of photo oxidation kinetics show a marked effect of PHB on the aging of blends of PHB and VAVAC. It is shown that irradiation of the films from the mixture of VAVAC and PHB (20%) lead to a change of pressure in the manometric cell associated with the absorption of oxygen. As for individual PHB the rate of photo oxidation of the blends of VAVAC and PHB essentially depends on the temperature. The activation energy of photo oxidation of 20 kJ/mol is determined from the slope of the line in the coordinates of the Arrhenius equation. This value is in a good agreement with the previously defined activation energy of photo oxidation of the individual PHB (16 kJ/mol [7]), but much higher than the activation energy of photo oxidation of the individual VAVAC (9 kJ/mol). Thus, the study of the kinetics of oxygen absorption during the photo oxidation of the mixtures of VAVAC and PHB suggests that the determining factor is the photo oxidation of PHB.

VAVAC absorbs light in the UV range (maximum at 275 nm and a shoulder at 313 nm). This absorption, in accordance with the published data [8], is characteristic also for polyvinyl alcohol and polyvinyl acetate. It is usually referred to carbonyl groups formed during the synthesis of polymers in the presence of acetaldehyde and oxygen impurities [8].

However, in some studies, this absorption is associated with polyconjugated structures, in particular the trienes [9].

Detailed spectrophotometric study performed on thin samples of VAVAC allowed us to obtain confirmation of the preliminary finding about the high photochemical activity of these groups made above.

As shown in Fig. 2, even at low irradiation time, ≤ 50 hours, the corresponding absorption is almost completely disappears. The transformation of these groups only to a small extent is accompanied by a broad structureless absorption in the near UV- and visible area attributed in accordance with the published data to polyconjugated structures (absorption maximum at 280–290 nm).

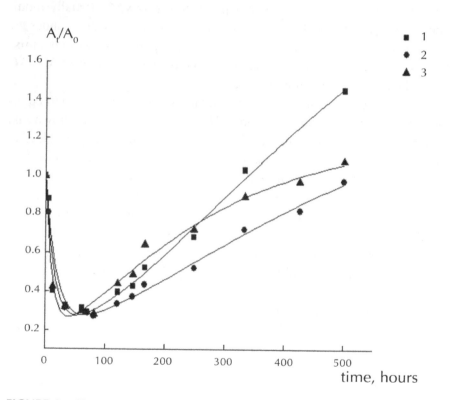

FIGURE 2 Changes in the relative optical density in the area of 357 (1), 312 (2) or 278 nm (3) during irradiation in air by the light of a mercury low pressure lamp DB-60 with λ=254 nm of the films with the thickness of 60 microns from VAVAC. Points – experimental data, curves – calculations by Eq. (1).The values of the parameters k_0, k_1 and k_2 for the dependence (1) (in brackets – for the dependence (2)) are 7.14×10^{-2}, (5.74×10^{-2}), 3.48×10^{-3} (3.17×10^{-3}) and 3.48×10^{-3} (5.01×10^{-3}) h^{-1}, and the values of the parameters a and b – 0.221 (0.224) and 2.38 (1.26), respectively.

Significant induction period of accumulation of secondary chromo-phore groups suggest the complexity of the process, which includes several stages. In the analyzed case, it can be represented as a sequence of two reactions of the 1st order, taking into account the conversion of the polymer PH units with forming of weakly absorbing photochemically active groups PAG, in turn, is converted into the polyconjugated structures PCS:

$$PH \xrightarrow{k_1} PAG \xrightarrow{k_2} PCS$$

It is necessary to take into account the spending of available in VAVAC chromophoric oxygen-containing groups OCG, formed during the synthesis of the copolymer and converted when exposed to light also by the reaction of the 1st order with the formation of not absorbing in the near UV- and visible area groups NAG:

$$OCG \xrightarrow{k_0} NAG$$

Then the change in the relative absorption (optical density) A/A_0 must comply with the following equation:

$$A/A_0 = a + (1 - a) \exp(-k_0 t) + b + [b/(k_2/k_1 - 1)]$$
$$\exp(-k_2 t) - \{b\, k_2/[\, k_1(k_2/k_1 - 1)]\} \exp(-k_1 t) \tag{1}$$

where a and b – parameters that take into account the absorption at "infinitely large" irradiation time t, and k_0, k_1 and k_2 – rate constants of the conversion of OCG, PH and PAG, respectively.

As seen in Fig. 2, the experimental data are well described by Eq. (1). It is essential that the constants k_1 and k_2 are almost two orders of magnitude smaller than k_0. Consequently, the primary chromophore groups OCG are not the immediate precursors of the polyconjugated structures PCS resulting from the prolonged irradiation, as well as photochemically active groups PAG serving as intermediate products.

Eq. (1) also describes the change in absorption during irradiation of thick films prepared from VAVAC, as well as from the mixtures of VAVAC and PHB (Fig. 1). In these cases, as evidenced by the kinetic analysis, the accumulation of products during prolonged photolysis is not directly related to the expenditure of the primary chromophore groups. It should be noted that the characteristic form of the absorption changes during irradiation of the films from VAVAC and from the mixtures of VAVAC and PHB does not depend on the spectral composition of light. The same types of dependencies were obtained by the action of polychromatic light of a mercury lamp, as well as during irradiation of the samples by the light of a xenon lamp in the machine "SUNTEST XLS+".

Expenditure rate (w) of the chromophore groups OCG, contained in VAVAC, essentially depends on the temperature (T). In Fig. 3, it is shown the dependence of w on the inverse temperature according to the Arrhenius equation

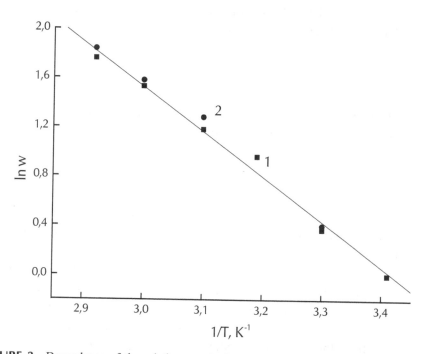

FIGURE 3 Dependence of the relative speed of expenditure of the functional groups OCG on the temperature during irradiation in air by polychromatic light of a mercury high pressure lamp of the films from a mixture of VAVAC and PHB (20%) (1) or from individual VAVAC (2). Points – experimental data, straight-line – calculations by Eq. (2) at the parameter values $w_0 = 1$, $T_0 = 293$ K, $E = 30.4$ kJ/mol.

$$\ln w = \ln w_0 - (E/R)(1/T - 1/T_0) \qquad (2)$$

where w_0 – speed at the "standard" temperature T_0 (in this case, $T_0 = 293$ K), E – activation energy, R – universal gas constant.

Defined by Eq. (2) the activation energy of the process in the range of 20–70°C is 30 kJ/mol. This is close to the activation energy of gaps and crosslinks in PVA in the range of 30–80°C (39 kJ/mol, irradiation by the light with $\lambda = 254$ nm in vacuum [10]). As might be expected on the basis of the above results, PHB has little effect on the rate of expenditure of chromophore groups during the photolysis of the mixtures of VAVAC and PHB. The activation energy of this process for mixtures containing 20% PHB does not differ from the corresponding value for the individual SVSVA (Fig. 3).

These results provide additional evidence that the effect of PHB on the photo-oxidation of VAVAC is not associated with changes in the activity of the chromophore groups of VAVAC in the areas of partial compatibility of the phases and near the interface boundary. Therefore it can be assumed, that this effect is largely due to the formation of active low-molecular radicals in PHB-phase during its photo oxidation. These radicals diffuse in the phase of VAVAC and initiate its oxidation. These radicals, apparently, are the radicals ·OH, because the water, along with CO_2, is the main product of photo oxidation of PHB [7].

Thus, the PHB as more easily oxidized component in the mixtures with VAVAC significantly reduces the induction period and increases the rate of accumulation of polyconjugated structures that lead to yellowing of the material under the action of UV-light. Corresponding the activation energy of photo-oxidation of the mixtures of VAVAC and PHB (20 kJ/mol) and the individual PHB based on partial data about the compatibility of the phases leads to the conclusion that the processes in PHB phase and transformation of PHB at the interface boundary are responsible for the acceleration of photo oxidation of VAVAC in the mixtures.

KEYWORDS

- **Photo-oxidation**
- **Polyconjugated structures**
- **Poly(3-hydroxybutyrate)**
- **Vinyl alcohol with vinyl acetate copolymer**

REFERENCES

1. Volova, T. G.; Sevastyanova, V. I.; Shishatskaya, E. I.; Polioksialkanoaty – biorazryshaemye polimery dlya mediziny. Krasnoyarsk: GK, "Platina", **2006**.
2. Fomin, V. A.; Gyzeev, V. V. Plast. Massy. **2001,** 2. 42.
3. Tertyshnaya, Yu. V.; Shibryaeva, E. S. Vysokomolek. Soed. A. **2004,** 46(7), 1205.
4. Olkhov, A. A.; Ivanov, V. B.; Vlasov, S. V.; Iordanskii, A. L. Plast. Massy. **1998,** 6, 19.
5. Mergaert, J.; Webb, A.; Anderson, C.; Wouters, A.; Swings, D. J. Appl. Environ. Microbiol. **1993,** 93(12), 3233.
6. Olkhov, A. A.; Iordanskii, A. L.; Shatalova, O. V.; Krivandin, A. V.; Vlasov, S. V. Vysokomolek. Soed. A. **2003,** 45(12). 2010
7. Ivanov, V. B.; Khavina, E. Yu.; Voinov, S. I.; Olkhov, A. A. Plast. Massy. **2007,** 1, 31.
8. Ranby, B.; Rabek, J. F.; Photodegradation, Photo-oxidation and Photostabilization of Polymers, Wiley, London, **1975**.
9. Geuskens, G.; Borsu, M.; David, C. Eur. Polym. J. **1972,** 8(7), 883.
10. Geuskens, G.; Borsu, M.; David, C. *Eur. Polym. J.* **1972,** *8(12)*, 1347.

CHAPTER 8

POLYOLEFIN CHLORINE-CONTAINING RUBBERS— PROPERTIES AND APPLICATIONS

YU. O. ANDRIASYAN, I. A. MIKHAYLOV, A. L. BELOUSOVA, G. E. ZAIKOV, A. E. KORNEV, and A. A. POPOV

CONTENTS

8.1 INTRODUCTION

The new alternative technology of obtaining chlorine-containing elastomers, based on solid phase (mechanochemical) halide modification was created taking into account current requirements. New chlorine-containing polyolefin cauotchoucs, obtained by given technology, showed yourself to good advantage in condition of rubber's production.

Based on historical data halide modification (HM) of high-molecular compound was carried out in 1859, natural rubber (NR) was exposed to modification and, in addition to that, NR was dissolved in perchloromethane, through which chlorine gas was run through. Modified NR is powder product with content of fixed chlorine not over 62–68% m., which didn't have properties of elastomer [1, 2]. Halide modification of NR may be referred to one of the first attempt of commitment of new properties to polymer with help of carrying out of chemical modification.

Nowadays, HM of polymers together with obtaining of halogen-containing polymers with help of synthesis is one of intensively developing direction in the field of obtaining chlorine-containing polymers. In result of carrying out of halide modification of polymers, which have technologically smoothly, large capacity industrial production, elastomer materials and composites are managed to obtain with wide complex of new specific properties: high adhesion, fire-, oil-, gasoline-, heat resistance, ozone resistance, incombustibility, resistance to influence of corrosive environments and microorganisms, high strength, gas permeability, etc.

Nowadays by world polymer industry was developed manufacture of those widespread polymers of halide modification, which has properties of elastomers such as: chlorosulfonated polyethylene (CSP), chlorinated polyethylene (CP), chlorinated and bromated butyl rubber (CBR, BBR) and chlorinated ethylene-propylene (CEP) and ethylene-propylene-diene cauotchoucs (EPDC) in small amount.

In the article we consider questions, concerning with obtaining and processing of halide modified chlorine-containing cauotchoucs as CBR and CEPC, which are prospective in terms of application in rubber industry. Perspectivity of their production and application consists in specific properties of these cauotchoucs (high gas permeability of CBR and high heat, ozone resistance of CEPDC). These properties are caused by structure of both initial (BR and EPDC) and chlorine-containing cauotchoucs (CBR and CEPDC).

Originally, before carrying out of halide modification of cauotchoucs BR and EPDC, attempts of rubber application based on these cauotchoucs were undertaken for purpose of items creation from elastomer materials, differing in high gas permeability and high heat-ozone resistance. In the process of properties study of rubber mixtures and rubbers from these cauotchoucs was found, that rubber mixtures had unsatisfactory characteristics by manufacturability of obtaining and processing. For the purpose of improvement of technological characteristics of rubber mixtures, attempts of combination of cauotchoucs BR and EPDC with diene cauotchoucs (natural rubbers, synthetic isoprene rubbers SIR-3, etc.) was undertaken. However, this combination didn't result in positive decision of given problem. If production and processing of rubber mixtures based on combined system of cauotchoucs with technological point of view didn't provoke difficulties, that creation of rubber items, which are able to use, is impossible. The reason is that if we combinate cauotchoucs, which differ in its unsaturation in case of application of sulfur vulcanization that resulted in absence of covulcanization between phases of combined cauotchoucs [3]. Thus, there were no unified, spatial, vulcanized network in rubbers based on combined system. In process of vulcanization took place redistribution of catalyst with help of diffusion and vulcanizing agent from the phase of cauotchouc with less unsaturation (BR, EPDC) to the cauotchouc's phase with high unsaturation (NR, SIR-3). Obtained rubbers are not satisfied with its strength and dynamic characteristics.

Many decisions of this problem were suggested, but the most effective was decision to add trace amount of halogen in macromolecular structure of cauotchouc with low unsaturation [4, 5]. It gave additional functionality to cauotchouc and therefore higher vulcanization rate. The optimum halogen content was when deterioration of initial cauotchoucs specific properties was not observed, and additionally capability of halide-containing cauotchouc to be covulcanized with high-unsaturated elastomers was gathered.

Historically, the most popular in tire industry was chlorine-butyl rubber (CBR). CEPDC cauotchouc had limited application, because required level of rubbers ozone resistance in rubber technology, traditionally, created by adding of chemical age resistors and antioxidants. Rubber ozone resistance and service time of the rubber item had the same duration. It's necessary to note, that this protect is inefficient for items with long period of service, because of exudation of age resistors and antioxidants from rubber. It is important to note, that the fraction of this items in common amount of output rubber products is very insignificant.

There were no problems on the first industrial production stage of halide modification of chlorine-containing cauotchoucs, the requirement in this cauotchoucs was growing, that was promoting to open new factories for production these cauotchoucs. However, it's necessary to note, that since realization of halide modification of natural rubber in 1859 almost nothing was changed in technology of obtaining of chlorine-containing cauotchoucs. This technology was preserved with some small changes until the present time. The meaning of given technology [6] or as it called by specialists "dissolved technology" is that, on the first stage, the polymer, which we want to modify, is dissolved in organic solvent. The concentration of solution should not exceed 10% from technological consideration. Then gaseous halogen, its chlorine or bromine, is leaked through obtained solution, then when planned content of halogen in polymer is reached, the process is suspended. Then are following the stages: detrainment of obtained chlorine-containing polymer, its washing and neutralization, then the stage of drying, packing and storage. As secondary process we can consider recuperation of solvent. All developments of this technology consisted in replacement of gaseous halogen with halogen-containing organic compounds, that didn't promote simplification of both technology and ecology of production process. Dissolved technology of obtaining of chlorine-containing cauotchoucs is multistage process, which in terms of current, strict ecology requirements doesn't stand up to scrunity.

Taking into account the above disadvantages of dissolved technology of obtaining of chlorine-containing cauotchoucs, alternative technology of obtaining of chlorine-containing cauotchoucs was developed and was offered in the end of ninetieth of past century by scientists' community and specialists of Moscow Academy of Fine Chemical Technology, Institute of biochemical physics, research and manufacturing association of firms "Polikrov" and The Moscow tire factory [7]. The differential characteristic of new technology is technological simplicity of carrying out obtaining process of chlorine-containing cauotchouc and its ecological safety.

The developed technology is based on solid phase (mechanochemical) halide modification of initial cauotchoucs by chlorine-containing organic compounds, which are environmentally safe in process of carrying out of halide modification. The developed technology has patent protection and opportunity to obtain both CBR and CEPDC, and others (saturate and unsaturated) cauotchoucs. Within the framework of newly developed technology is assimilated experimental-industrial output of cauotchoucs CBR-2,5 and CEPDC-2,0 (the number shows content of fixed chlorine in cauotchouc).

8.2 RESULTS AND DISCUSSION

Research-industrial testing of cauotchouc CBR-2,5 in rubber formula of radial tires inner lining, tubeless construction, was carried out on the Moscow tire factory. The point of carrying out of investigations was to substitute serially used rubbers of inner lining of chlorine-containing cauotchouc HT-1066 (produced by USA) for cauotchouc CBR-2,5. Conducted investigations showed, that production and processing of rubber mixtures with new cauotchouc CBR-2,5 on technological equipment created no problems. Plasto-elastic, physical-mechanical and some specific properties of serial and experiment rubber mixtures and their vulcanizates, containing cauotchouc CBR-2,5 were studied. The results of investigations are showed in Table 1.

TABLE 1   Properties of rubber mixtures and rubbers for radial tires' inner lining with application of serial chlorine-butyl rubber CBR HT-1066 and CBR-2,5.

Index	HT-1066	CBR-2,5
Plasticity	0.37	0.40
Cohesion strength, MPa	3.49	3.45
Mooney viscosity (100°C)	58.5	66.0

TABLE 1   *(Continued)*

Plastometer "Faerston" tests		
Flow time of rubber mixture, s.	25.8	16.2
Shrinkage, %	62.0	58.5
Monsanto rheometer tests		
Rotational moment, N*m		
Min	9.0	9,8
Max	16.0	24,5
Initial time of vulcanization, min	4.4	9,3
Vulcanization rate, %/min	7.9	9,4
Optimum vulcanization time, min	17.0	20,0
Physical-mechanical indexes		
Conventional tensile strength 300%, MPa	4.2	6.9
Conventional tensile strength, MPa	10.5	10.0
Conventional breaking elongation, %	650	550
Tear resistance, kN/m	31	39
Gas permeability (to hydrogen), $l/(m^2{*}d)$	0.49	0.52

We can see from the Table 1, that experimental and serial rubbers almost didn't differ in plasticity, Mooney viscosity, and cohesion strength.

Test of rubber mixtures on plastometer "Faerston" found higher fluidity of experimental rubber with cauotchouc CDR-2,5.

Study of vulcanized characteristics of the rubber mixtures on Monsanto rheometer showed, that experimental mixtures with cauotchouc CDR-2,5 excel serial mixtures almost in two times in the initial vulcanization time and have higher vulcanization rate in basic period, that is very important with technological point of view.

Study of physical-mechanical characteristics of the rubbers showed, that experimental rubber much excel serial in tensile strength 300%, but there are no differences between experimental and serial rubbers in strength, conventional breaking elongation and tear strength.

The values of gas permeability (to hydrogen) of experimental and serial rubbers are the similar.

Thus, in the Table we showed, that new chlorine-containing butyl cauotchouc CBR-2,5 satisfies the requirements by their characteristics, demanded on halogen-containing butyl cauotchoucs, used in rubber production of inner lining.

The next stage of our investigations was to study opportunities of application of new chlorine-containing cauotchouc CEPDC-2 in the formulas of rubbers for sidewall of radial tires and rubbers for production of diaphragm press.

As we know, sidewalls' rubber is exposed deformations in process of service, which is the reason of intensive heat emission. Increased temperature promotes premature heat and ozone ageing of rubbers of tires sidewalls. Traditionally chemically synthesized antioxidants and age resistors are mixed in rubber formula for protection of sidewall rubber from heat and ozone ageing [8]. The "bleeding" of protectors takes place in process of service, because they don't bind chemically with elastomer matrix; all this reasons promote premature ageing and destruction of sidewalls. Considering that part of tires can be reconstructed after service period, it will be very practically advantageous to increase heat and ozone resistance of sidewalls by adding of protection component, which can build into elastomer matrix with help of chemical links. The function of that component can make new chlorine-containing cauotchouc CEPDC-2, because we know, that it has capability to covulcanize with high-unsaturated cauotchoucs, composing on rubber formula for sidewalls. It is well known, that adding 20–30 mass part of cauotchouc CEPDC-2 is enough for increase of ozone resistance of rubber from diene cauotchoucs [9].

In this case we studied the opportunity of application of cauotchouc CEPDC-2 in rubber formula for sidewalls of radial tires, elastomeric part of which has diene cauotchoucs SIR-3 and CDR in ratio (50:50). The ratio of cauotchoucs SIR-3:CDR:CEPDC-2 was 50:20:30 and 50:30:20 in experimental rubber. Chemical antioxidants weren't added in experimental rubber mixtures.

We established, that production and processing of rubber mixtures with cauotchouc CEPDC didn't have difficulties on technological equipments. We studied plasto-elastic, physical-mechanical and some specific properties of serial and experiment rubbers. Experimental data are showed in Table 2.

TABLE 2 Properties of studied serial and experimental rubber mixtures and rubbers, based on cauotchoucs SIR-3, CDR, CEPDC-2.

Index	Serial rubber*	Experimental rubber	
		1**	2***
Mooney viscosity (120°C)	43	45	47
Plasticity	0.44	0.44	0.36
Conventional modulus at 300%, MPa	3.7	7.2	6.0
Conventional tensile strength, MPa	15.7	20.5	18.4
Conventional breaking elongation, %	770	600	610
Conventional permanent tension elongation, %	14	15	13
Coefficient of heat ageing (100°C, 72 h)			
At strength	0.56	0.85	0.82
At conventional elongation	0.63	0.92	0.91
Coefficient of ozone resistance of dynamic tests ($\varepsilon=20\%$)	0.52	0.95	0.92
TM-2 hardness	56	60	62
Rebound elasticity, %			

TABLE 2   *(Continued)*

Under 20⁰C	41	42	44
Under 100⁰C	47	50	52
Crazing strength, th. cycle	>252	>252	>252
Dynamic repeated tension durability, th. cycle	>50	>50	>50

*Based on cauotchoucs SIR-3 - CDR (50:50),

**Based on cauotchoucs SIR-3 – CDR – CEPDC-2 (50:30:20),

***Based on cauotchoucs SIR-3 – CDR – CEPDC-2 (50:20:30).

We can see from the Table 2, that plasto-elastic characteristics of serial and experimental rubbers have close values, serial rubbers have conventional modulus at 300% twice higher as experimental ones and have higher conventional tensile strength and hardness. The value of rebound elasticity, crazing strength and dynamic repeated tension durability of experimental and serial rubbers are very similar. It should be noted, that serial rubbers have heat resistance and ozone resistance doubles that serial rubbers, containing antioxidants.

Thus, the investigations showed, that new chlorine-containing cauotchouc CEPDC-2 in rubber formulas for tires sidewalls can be used as polymer antioxidant.

The practice states that the main reasons of breakdown of diaphragm press are the low capacity to elastic recovery of rubbers based on butyl cauotchoucs, leading to "treading out" of diaphragm, and high extent of "tar value" of diaphragm work surface. To eliminate these disadvantages we studied opportunity of substitution of cauotchouc SEPC-60 in formulas of serial rubbers (resin curing) for diaphragm for new chlorine-containing ethylene-propylene-diene cauotchoucs CEPDC-2.

Cauotchoucs BR-1675 and SEPC-60 in ratio (85:15) compose formula of serial rubbers for diaphragm production, in experimental rubber SEPC-60 was substituted for similar amount of CEPDC-2. It is well known, that chlorine-containing compounds have the ability to activate resin curing of butyl cauotchouc, which is the main elastomer component of membraneous rubbers [1].

The investigations showed, that there were no difficulties in production and processing of rubber mixtures with cauotchouc CEPDC-2 on technological equipments.

We studied plasto-elastic, physical-mechanical and some specific characteristics of serial and experiment rubber mixtures and rubbers. Experimental data are showed in Table 3.

TABLE 3   Properties of serial and experimental rubber mixtures and rubbers for production of shaper-vulcanization.

Index	Serial rubber	Experimental rubber
Plasticity	0.41	0.42
Mooney viscosity (140°C)	37	36
Conventional modulus at 300%, MPa	5.0	6.0
Conventional tensile strength, MPa	10.2	12.6
Conventional breaking elongation, %	620	600
Conventional permanent tension elongation, %	34	20
Tear resistance, kN/m	60	63
Coefficient of strength heat ageing (180°C, 24 h)	0.6	0.6
Coefficient of strength temperature resistance under 100°C	0.7	0.62
TM-2 hardness	74	78
Dynamic repeated tension durability (ε_{dyn}=50%; ε_{stat}=37.5%), th. cycle	42	>50
Creep (160°C, 24h, 0.3 MPa), mm	119	53
Rebound elasticity*, %	13/18	18/32
Rebound elasticity*, after ageing, %	16/28	18/30
Tar value, %	1.2	0.6

*In numerator under 20°C, in denominator under 100°C.

We can see from the Table 3, that plasticity and Mooney viscosity of experimental and serial rubber mixtures have close values. The values of conventional modulus at 300%, conventional tensile strength, tear resistance, hardness, rebound elasticity and dynamic durability of experimental rubbers are higher than serial ones. It should be noted, that experimental rubber has lower by half conventional permanent tension elongation in comparison with serial rubber, although the values of conventional tensile elongation are the similar; "tar value" and creep under 160⁰C are lower (in two times and more, than in two times respectively).

8.3   CONCLUSION

Thus, we can make conclusion, based on obtained results that application of new chlorine-containing cauotchouc CEPDC-2 in rubber formula for production of press diaphragm will permit to increase the diaphragm service time. It should be noted, that developed new technology of obtaining of chlorine-containing cauotchoucs permits to manufacture competitive chlorine-containing polyolefine cauotchoucs CBR-2,5 and CEPDC-2. As showed the investigations of cauotchouc CBR-2,5, well recommended yourself in conditions of rubber production and cauotchouc CEPDC-2, which didn't have analogues on synthetic cauotchoucs market, can be used as polymeric antioxidant in rubbers based on diene cauotchoucs.

KEYWORDS

- Chlorinated ethylene-propylene
- Chlorosulfonated polyethylene
- Ethylene-propylene-diene cauotchoucs
- Halide modification
- Mechanochemical
- Perchloromethane
- Polyolefin cauotchoucs

REFERENCES

1. Doncov, A. A.; Lozovik, G.; Novizkaya, S. P.; Chlorined polymers. M. "Chem.", **1979**, 232.
2. Ronkin, G. M. Current state of production and application of chlorine polyolefine, M.: NIITEChem, **1979**, 81.
3. Chirkova, N. V.; Zaxarov, N. D.; Orexov, S. V. Rubber mixtures based on combination of cauotchoucs. M., **1974**. 62.
4. Morrissey, R. T. Halogenation of Ethulene Propylene Diene Rubbers. Rubber Chem. Technol. **1971**, *44(4)*, 1025–1042.
5. Ronkin, G. M. Investigation of chloring ethylene-propylene-diene copolymers process and properties of obtained modifications. Ronkin, G. M. et al. Manufacturer of SC. **1981**, *6*, 8–11.
6. Ronkin, G. M. Chlorosulfonated polyethylene. M., **1977**, 101.
7. Andriasyan, Y. O. Elastomeric materials based on cauotchoucs, exposed mechanochemical halide modification. 05.17.06. M., **2004**, 362.
8. Ragulin, V. V. Production if pneumatic tires. M.: Chem., **1965**, 504.
9. Andriasyan, Y. O. Investigation of properties of rubber mixtures and vulcanizates based on combined system of unsaturated cauotchoucs with halogenated ethylene-propylene cauotchoucs. 05.17.12. M., **1981**, 212.

CHAPTER 9

ADHESION CHARACTERISTICS OF ETHYLENE COPOLYMERS

N. E. TEMNIKOVA, S. N. RUSANOVA, S. YU. SOFINA,
O. V. STOYANOV, R. M. GARIPOV, A. E. CHALYKH,
V. K. GERASIMOV, and G. E. ZAIKOVPETRONIS

CONTENTS

9.1   INTRODUCTION

The influence of aminoalkoxy- and glycidoxyalkoxysilanes on the strength of adhesive contact modified polymer-steel (aluminium, polyethyleneterephthalate) for acetate ethylene copolymers was studied.

Polyethylene and copolymers of ethylene with vinylacetate (EVA) and from the beginning of the 21st century copolymers of ethylene with vinyl acetate and maleic anhydride (SEVAMA) are commonly used as the main ingredients of hot melt adhesives, which determine the cohesive strength, heat resistance, processing conditions and the use of the composition. It is known that aminoalkoxysilanes and organosilanes containing glycidoxy groups are widely used in fiberglass and paint industries improve the adhesion of different polymers and coatings (acrylic, alkyd, polyester, polyurethane) to inorganic substrates (glass, aluminum, steel, etc.). Therefore increase of the adhesive strength of the compositions based on polyolefins and these organosilanes to various substrates (steel, aluminum) is to be expected.

Earlier [1–5], we studied the changes in the chemical structure of ethylene copolymers during organosilane modification and the effect of organosilicon compounds on the structural, mechanical, operational and technological characteristics of the modified copolymers. The purpose of this work was to study the adhesion characteristics of ethylene copolymers modified by aminoalkoxy- and glycidoxyalkoxysilanes.

9.2   SUBJECTS AND METHODS

Copolymers of ethylene with vinyl acetate (EVA) Evatane 20–20 and Evatane 28–05; copolymers of ethylene with vinyl acetate and maleic anhydride (SEVAMA) grades Orevac 9305 and Orevac 9707 were used as the objects of the study. The main characteristics of the polymers are shown in Table 1. Modifier – (3-glycidoxypropil)trimethoxysilane. The reaction mixing of copolymers with modifiers was conducted on laboratory roll mills at a rotational speed of rolls 12.5 m/min and friction 1:1.2 for 10 minutes in the range of 100–120°C. Modifier content was varied in the range of 0–10 wt%. Adhesive joint strength of the samples was evaluated by flaking method the day after forming an adhesive compound in accordance with GOST 411–77 (to metal substrates) and GOST 2896.1–91 (to polyethyleneterephthalate). Metal substrates: steel 3 (St3) and aluminum AMG-1.

TABLE 1 Characteristics of ethylene copolymers.

Brand of EVA	EVATANE 20–20	EVATANE 28–05	Orevac 9305	Orevac 9307
Unit designation	EVA20	EVA27	SEVAMA26	SEVAMA13
VA content, %	19–21	27–29	26–30	12–14
MA content, %	–	–	1.5	1.5
Melt flow rate, r/10 min, t = 190°C, load 5 kg	59.29	18.16	92.77	26.76
Melt flow rate, r/10 min, t = 125°C, load 2.16 kg	2.67	0.76	14.72	1.02
Density, kg/m^3	938	950	951	939
Breaking stress at tension, MPa	14	23.88	7	19.5
Elongation at break, %	740	800	760	760
Elastic modulus, Mpa	31	17	7	61

9.3 RESULTS AND DISCUSSION

The interaction of functional groups of adhesive and substrate was considered in many studies, but the optimal content of active groups in the adhesive is often selected empirically, because in most cases there is no proportionality between the adhesive strength and the content of the functional groups in adhesive. According to Ref. [6], this dependence is often extreme, or with the increase of the content of the functional groups adhesive strength reaches a certain limit and no longer increases.

It was found that the introduction of 2% by mass of AGM-9 increases the adhesion strength to steel (St3) for EVA27 in 4 times and for SEVA-MA26 in 1.75 times. The further introduction of the modifying agent does not lead to an increase in the strength of the adhesive compound (Fig. 1).

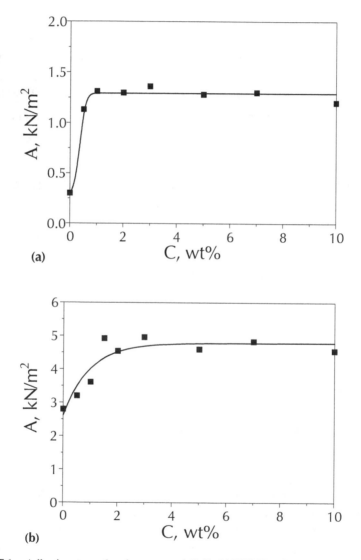

FIGURE 1 Adhesive strength polymer – steel (St3): (a) EVA27-AGM-9; (b) SEVAMA26-AGM-9. Condition of formation 160°C, 10 min.

It was also found that the introduction of up to 0.5 wt% DAS and AGM-9 increases the adhesion strength to aluminum (AMG-1) for the copolymers EVA20-AGM-9 in 1.75 times and for SEVAMA26-DAS in 14 times. The further increase of the modifying agent concentration reduces the adhesion strength (Fig. 2) relatively to the maximum value.

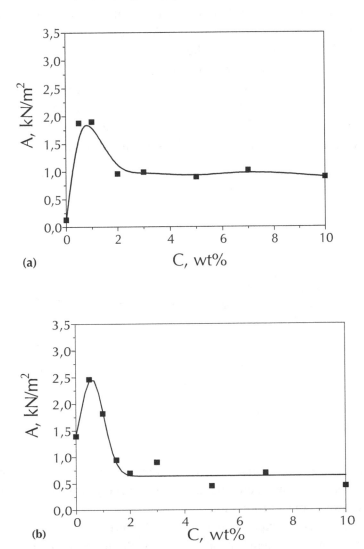

FIGURE 2   Adhesive strength polymer – aluminum (AMG-1): (a) EVA20-AGM-9; (b) SEVAMA26-DAS. Condition of formation 160°C, 10 min.

For terpolymers destruction of the system adhesive – metal with the modifier content up to 2% has the mixed nature; a bundle of the adhesive coating takes place, because adhesive joint strength of the system polymer – metal is higher than the strength of the adhesive. In this case, the adhesive strength of the modified and non-modified terpolymers is significantly higher than the adhesive strength of the double copolymers, due to the presence of maleic anhydride in SEVAMA. At a modifier concentration of more than 2% the gap is purely adhesive.

Modification of ethylene with vinyl acetate copolymers by glycidoxyalkoxysilane has no effect on the strength of the adhesive contact ethylene copolymer – PET. However, the introduction of 1.5 wt% of this additive in SEVAMA leads to a significant (in 3.4 times) increase in adhesion strength of the system, and the destruction of the substrate is observed (Fig. 3).

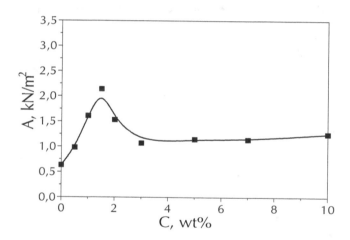

FIGURE 3   Adhesive strength polymer – PET: SEVAMA26-GS. Condition of formation 160°C, 10 min.

Adhesive strength of the system polymer – substrate depends on the nature of the contacting materials (their surface energy) and the conditions of the adhesive contact formation.

Table 2 shows the experimental data on the surface energy and its polar and dispersion components for all grades of the tested copolymers. The surface energy of the initial copolymers is determined mainly by the dispersion component.

TABLE 2   Energy characteristics of the copolymers.

Grade	γ, mJ/m²	γ^D, mJ/m²	γ^P, mJ/m²
EVA20	36	25.1	10.9
EVA27	38.6	25	13.6
SEVAMA13	40.4	27.9	12.5
SEVAMA26	37.4	28.4	9

These data were applied on the generalized temperature-concentration diagram of the surface energy of EVA (Fig. 4), taken from Ref. [7], which includes both the operating conditions and testing of adhesive joints (curve 1), and also the conditions of the adhesive joints formation (curve 2). Traditionally [8], dependencies of the surface energy from the composition of the random copolymers describe by the simple additive function – the dotted lines. As can be seen in Fig. 4, the surface energies of the copolymers are close to additive values.

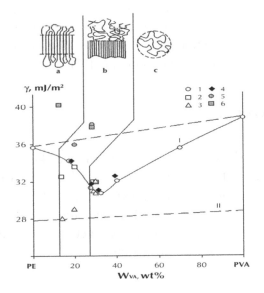

FIGURE 4   Dependence of the surface energy from the composition of EVA copolymers. The samples were formed in contact with: air – 1 [9], 2, 4 [8], PTFE – 3 [10], 5 – EVA (PET), 6 – SEVAMA (PET).

In Ref. [10], it was shown that the total surface energy of the copolymers is a function of the substrate surface nature, in contact with which the surfaces of the copolymers samples were formed. Typically, the surface energy of the samples obtained on high-energy surfaces (steel, glass, aluminum, PET), is significantly higher than the surface energy of the samples prepared in contact with low-energy surfaces (PTFE, its copolymers, PE), and in the air. Since the formation of the samples was carried out on PET, this explains the higher values of the surface energy for the studied copolymers, relative to the copolymers formed in contact with low-energy surfaces.

Thus, on the basis of these results in Ref. [7] concluded that, when exposed to high-energy substrates in the surface layers of the copolymers conformational changes take place, that result in changes of vinylacetate units concentration, on the one hand, and in changes in the packing density of macromolecules segments on the other hand. For high-energy substrates the number of vinylacetate units in the surface layers of the copolymers is close to their number, characteristic for the surface of PVA. Formation of the adhesive compound modified copolymer – substrate was carried out in contact with the high-energy substrates. Unreacted groups of VA along with grafted siloxane concentrate on the composition surface. Both of these factors, complementing each other, lead to increase of adhesion between composition and steel (aluminum, PET).

9.4   CONCLUSION

Modification of ethylene copolymers by aminoalkoxy-, glycidoxyalkoxysilanes increases the strength of the adhesive contact both to metals (steel, aluminum) and to polar polymers (polyethyleneterephthalate). Therefore, these materials having enhanced deformation, strength and adhesive properties may be used in the production of multilayer polymer films, including metallized polymer films.

KEYWORDS

- Ethylene with vinylacetate
- Glycidoxyalkoxysilanes
- Polyethyleneterephthalate
- Polymer films

REFERENCES

1. Rusanova, S. N. Diss. tekhn. nauk: 02.00.16 Rusanova Svetlana Nikolaevna. Kazan: Kazanskii gos. tekhnol. un-t, **2000,** 119.
2. Temnikova, N. E.; Rusanova, S. N.; Tafeeva, Yu. S.; Sofina, S. Yu.; Stoyanov, O. V. Vliyanie aminosoderzshashego modificatora na svoistva sopolimerov etilena. Klei. Germetiki. Tekhnologii. **2012,** *4,* 32–48.
3. Rusanova, S. N.; Temnikova, N. E.; Mukhamedzyanova, E. R.; Stoyanov, O. V. Modifikatsiya sopolimerov etilena aminotrialkoksisilanom. Vestnik Kazanskogo tekhnologicheskogo universiteta. **2010,** *9,* 353–355.
4. Rusanova, S. N.; Temnikova, N. E.; Stoyanov, O. V.; Gerasimov, V. K.; Chalykh, A. E. IK-spektroskopicheskoe issledovanie vzaimodeistviya glitsidoksisilana i sopolimerov etilena. Vestnik Kazanskogo tekhnologicheskogo universiteta. **2012,** *22,* 95–97.
5. Temnikova, N. E.; Rusanova, S. N.; Tafeeva, Yu. S.; Stoyanov, O. V. Issledovanie modificatsii sopolimerov etilena aminosilanami metodom IK-spektroskopii NPVO. Vestnik Kazanskogo tekhnologicheskogo universiteta. **2011,** *19,* 112–125.
6. Currat, C. Silane croslinker insulation for medium voltage power cables. Currat C. *Wire J. Int.* **1984,** *17(6),* 60–65.
7. Gladkikh, Yu. Yu. Deformatsionno-prochnostnye i adgezionnye svoistva sopolimerov etilena i vinilatsetata: Diss. kand. khim. nauk. Gladkikh Yuliya Yurievna. – Moskva: IFKhE RAN, **2012,** 156.
8. Vu, S. Mezshfaznaya energiya, struktura poverkhnostei i adgeziya mezshdy polimerami. v sb.: Polimernye smesi pod red. Pola D. i Niumena S. M.: Mir, **1981,** *3(1),* S.282–332.
9. Nose, T. Theory of Polymer Liquids and Glasses, A Hole. *Polymer J.* **1972,** *3(1),* 1–11.
10. Balashova, E. V. Vliyanie predistorii na poverkhnostnye svoistva polimerov v razlichnykh fazovykh sostoyaniyakh: Diss. kand. khim. nauk. Balashova Elena Vladimirovna. Moskva: IFKhE RAN, **2003,** 155.

CHAPTER 10

OXIDATION OF FIBRINOGEN BY OZONE—EFFECT OF DIHYDROQUERCETIN AND CYCLODEXTRIN INCLUSION COMPLEX WITH THE NEW DIHYDROQUERCETIN DERSIVATIVE

V. S. ROGOVSKY, T. M. ARZAMASOVA, M. A. ROSENFELD,
M. L. KONSTANTINOVA, V. B. LEONOVA, S. D. RAZUMOVSKY,
G. E. ZAIKOV, A. I. MATYOUSHIN, N. L. SHIMANOVSKY,
A. M. KOROTEEV, S. E. MOSYUROV, M. P. KOROTEEV,
T. S. KUHAREVA, and E. E. NIFANTIEV

CONTENTS

10.1 INTRODUCTION

Dihydroquercetin (taxifolin, DHQ) is a natural flavonoid, which possesses antioxidant activity and other pharmacological properties (anti-inflammatory, anti-atherosclerotic, etc.). Dihydroquercetin is hydrophobic compound, that's why it can't be administered intravenously, also its oral bioavailability is reduced. Recently, many new dihydroquercetin derivatives were synthesized, including water-soluble forms (cyclodextrin inclusion complexes with dihydroquercetin derivatives). In addition to the protective effect of antioxidants against lipid peroxidation, increasing attention is paid to the possibilities of antioxidants including dihydroquercetin to prevent an oxidation of proteins. Fibrinogen is more susceptible to oxidation than most other plasma proteins.

In this chapter, we study the ability of water-soluble dihydroquercetin derivative to inhibit ozone oxidative modification of fibrinogen. Assessment of functional activity of fibrinogen both before and after oxidation with ozone was conducted by determining the time of formation of a fibrin clot after addition of thrombin to fibrinogen solution. Recently, we have evaluated the ability of native dihydroquercetin to inhibit ozone oxidation of fibrinogen. In this chapter, we found that the new water-soluble dihydroquercetin derivative is more potent in preventing the oxidative modification of fibrinogen in comparison with native dihydroquercetin.

Oxidative modification of fibrinogen inhibits thrombin-catalyzed clot formation [1], inhibits platelet adhesion and aggregation [2], enhances the activity of tissue plasminogen activator, cause negative effect on hemorheological parameters [3], and is actively involved in the process of inflammation of the vascular wall. Also oxidized fibrinogen is able to induce increased production of interleukin-8 in primary culture of human endothelial cells, which causes chemotaxis of immune cells [4–6]. Oxidation of fibrinogen causes an *imbalance in the blood coagulation* system. In blood plasma oxidized fibrinogen can form clots with impaired structure, which is more resistant to fibrinolysis [7]. This fact can lead to ischemic complications of increased oxidative stress. Most of antioxidants are hydrophobic compounds that's why they can't be administered intravenously also their oral bioavailability is reduced. Thus the learning of ability of various compounds (especially with increased water-solubility) to prevent oxidative damage to blood plasma proteins particularly to fibrinogen is relevant. The objective of this work is to investigate the ability of *the new* water-soluble dihydroquercetin derivative (KN-14–CD) to inhibit the o*xidative* modification of *fibrinogen.*

Flavonoids are the class of plant polyphenols with a wide spectrum of biological activities. Flavonoids protect plants against different kinds of stress. They act as a UV-filter, function as a signal molecules, antimicrobial defensive compounds, etc. [8].

Dihydroquercetin (taxifolin, DHQ) is the natural flavonoid, which possesses antioxidant activity and lots of other pharmacological properties (anti-inflammatory, anti-atherosclerotic, etc.) [9]. Compounds with an antioxidant activity are increasingly used in the treatment of diseases in which the induction of oxidative stress is one of the key elements of pathogenesis. Such diseases include ischemic attacks, ischemic heart disease, vasculitis and many others.

Dihydroquercetin is hydrophobic compound, that's why it can't be administered intravenously, also its oral bioavailability is reduced. Recently many new dihydroquercetin derivatives were synthesized including water-soluble forms (cyclodextrin inclusion complexes with dihydroquercetin derivatives).

Cyclodextrins are cyclic oligosaccharides. Their molecules have a hydrophilic outside, which can dissolve in water, and hydrophobic cavity. As a result of this cavity, cyclodextrins are able to form inclusion complexes with a wide variety of hydrophobic guest molecules. This can cause an improved bioavailability and increased pharmacological effects of guest molecules [10].

The new aminomethylated dihydroquercetin derivative (KN-14) is even more hydrophobic compound than native dihydroquercetin. That's why KN-14 forms more stable inclusion complex with cyclodextrin that cause increased water-solubility of resulting compound.

In addition to the protective effect of antioxidants against damage to cell membranes increasing attention is paid to the possibilities of antioxidants to prevent oxidation of proteins. It is generally recognized today that many amino acid residues of proteins are susceptible to oxidation by reactive oxygen species (ROS). Free-radical oxidation of proteins may be accompanied by the splitting of the polypeptide chains, a modification of amino acids and transformation of proteins to compounds which are highly sensitive to proteolytic degradation [11]. Oxide-modified proteins are accumulated in the body with age, under oxidative stress and a number of diseases, particularly diabetes mellitus [12–14]. Fibrinogen is more susceptible to oxidation than most other plasma proteins (2). Many diseases cause increased oxidation of fibrinogen, which significantly reduces its functional activity. It is shown that fibrinogen is at least 20 times more sensitive to oxidative modification in comparison with other major plasma proteins: albumin, immunoglobulins, transferrin, ceruloplasmin [15, 16]. That leads to the formation of oxidized forms of the protein, different from the native form in the chemical composition and the structural organization. Formation of non-covalent bonds enables assembling of *macromolecular fibrinogen clusters* [16]. Oxidative modification of fibrinogen inhibits thrombin-catalyzed clot formation [1], inhibits platelet adhesion and aggregation [2], enhances the activity of tissue plasminogen activator, cause negative effect on hemorheological parameters [3], and is actively involved in the process of inflammation of the vascular wall. Also oxidized fibrinogen is able to induce increased production of interleukin-8 in primary culture of human endothelial cells, which causes chemotaxis of immune cells [4–6]. In blood plasma oxidized fibrinogen can form clots with an impaired structure, reduced clot porosity and delayed fibrinolysis in comparison with healthy patients [7, 17].

This phenomenon can lead to an ischemic complication of increased oxidative stress. Thus the study of ability of various compounds (especially with increased water-solubility) to prevent oxidative damage to blood plasma proteins particularly to fibrinogen is relevant. The objective of this work was to investigate the ability of *the new* water-soluble dihydroquercetin derivative (KN-14-CD) to inhibit the *oxidative* modification of *fibrinogen*.

10.2 EXPERIMENTAL

In this work, we carried out the oxidation of fibrinogen solution under standard conditions [4] and in the presence of cyclodextrin inclusion complex with the new aminomethylated dihydroquercetin derivative (KN-14-6-[(diFluorine-butylamino)methyl]dihydroquercetin). We used 2-hydroxypropyl-β-cyclodextrin, because it can be appropriable for intravenous administration [10, 18].

In the study, we used bovine fibrinogen obtained from purified citrated blood plasma [13]. Fibrinogen was transferred to 0.06 M phosphate buffer (pH 7.4) containing 0.15 M NaCl using Sephadex G-25 gel filtration.

To determine the effect of dihydroquercetin on the oxidation of fibrinogen we compared the UV spectra of test solutions before and after oxidation. Functional activity of fibrinogen was assessed by controlling its ability to form a fibrin clot after addition of thrombin. In the studying of the effect of KN14-CD on the oxidation of fibrinogen we used ozone as the oxidant. Ozone was chosen for the following reasons. This natural agent belongs to the reactive oxygen species, and it is convenient to carry out research in model systems with ozone.

Its half-life in aqueous solution in the presence of oxidation substrates does not exceed 1–2 min. The oxidation state of the object is strictly regulated, because the amount of ozone that reacts with a reducing agent is precisely measured spectrophotometrically at 254 nm. Ozonation of fibrinogen solution was carried out in a reactor (volume 3.3×10^{-3} l) equipped with quartz windows; ozone-oxygen mixture was blown through a free volume of the reactor. The amount of ozone in the reactor was 2.0×10^{-7} M. Ozone was synthesized in a discharge device by passing a stream of oxygen between the electrodes. The UV spectra of proteins before and after ozonation were recorded on SF-2000 spectrophotometer (Russia) in a 1 cm quartz cells.

To study kinetics of absorption of ozone by KN-14-CD we used bubbling method in which ozone is passed through a layer of a solution containing the test compound, fixing the change in ozone concentration before and after passing through the solution. Assessment of functional activity of fibrinogen both before and after oxidation with ozone was conducted by determining the time of formation of a fibrin clot after addition of thrombin ("Roche", France) in a solution of fibrinogen. Thrombin solution (0.05 ml, 2.5 units NIH) was added to 0.3 ml test sample.

10.3   RESULTS AND DISCUSSION

Bubbling method of ozone oxidation was used to study kinetics of the interaction of ozone with KN-14-CD. Concentration of KN-14-CD was 6.25×10^{-7} mol/l. Kinetics of ozone absorption by KN-14-CD is shown in Fig. 1.

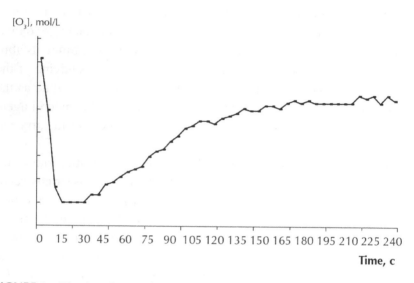

FIGURE 1   Kinetics of ozone absorption by KN-14-CD.

According to the data presented in Fig. 1 using Eq. (1), we have calculated the rate constant for the interaction of ozone with KN-14-CD and stoichiometric coefficient of this reaction, which is determined by the ratio of the amount of absorbed ozone to the amount of substance reacted. Stoichiometric coefficient of the reaction of ozone with KN-14-CD is 3 (the same as DHQ has) [19]. The rate constant of the interaction of ozone with KN-14 is 5.0×10^3 l/mol.s. According to our previous study rate constant of the interaction of ozone with DHQ is 1.0×10^3 l/mol.s. That indicates a greater reaction rate of KH-14-CD with ozone, and hence a higher antioxidant properties of KH-14.

The equation for determining the rate constant for the interaction of ozone with KN-14 [20].

$$k = W \times ([O_3]_0 - [O_3]_{gas})/\alpha[O_3]_{gas}[reagent] \qquad (1)$$

where, W – flow rate of the gas mixture (50–200 l/min); α – solubility of ozone in water (0.32, experimental determination); $[O_3]_0$ – initial concentration of ozone; $[O_3]_{gas}$ – concentration of ozone, which leaves the reactor; [reagent] – initial concentration of reagents.

Before study of combined fibrinogen and KN-14-CD oxidation by ozone we previously studied the interaction of native fibrinogen with ozone. For this purpose, we carried out the oxidation of fibrinogen in a closed reactor for 10 min. The intensity and character of the oxidation process were determined using UV spectra of unoxidized (native) and oxidized fibrinogen with cyclodextrin. Figure 2 shows the spectra of native and oxidized fibrinogen. Ozone oxidation of fibrinogen was assessed by the change in optical density at a wavelength of 280 nm (absorbance of the amino acid residues containing phenolic groups, such as tyrosine, phenylalanine, and unsaturated residues of tryptophan and histidine).

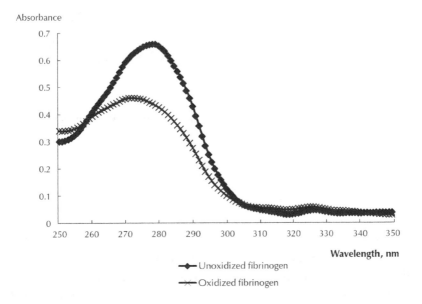

FIGURE 2   UV spectra of fibrinogen (1.87 mg/ml) before and after ozone oxidation.

The spectrum of oxidized fibrinogen shows that ozone oxidation of the protein decreases the amplitude of the fibrinogen absorption peak, and the peak shifts to shorter wavelengths. Also we can note the appearance of a new minor absorption peak at 325 nm, which may be due to the formation of quinoid structures of the phenoxyl and imidazole core of amino acid residues under the influence of ozone [19].

Next, we studied the ozone oxidation of fibrinogen in the presence of KN-14-CD (Fig. 3).

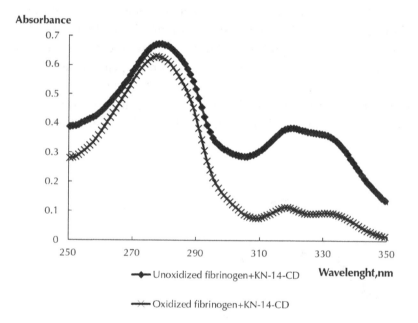

FIGURE 3   UV spectra of fibrinogen (1.87 mg/ml) in the presence of dihydroquercetin $(1.0\times10^{-5}$ M) before and after ozone oxidation.

By comparing the spectra presented in Figs. 2 and 3, we see that in the case of the oxidation of fibrinogen in the presence of KN-14-CD absorption peak of fibrinogen is decreased by 5%, whereas in the absence of KN-14-CD we can see a significant decrease in absorption peak of fibrinogen by 30%. Thus, we can conclude that KN-14-CD in this case has a protective effect against ozone oxidation of fibrinogen, because it has ability to react with ozone and therefore prevent the oxidative degradation of fibrinogen.

According to our previous study (which was confirmed in current work), KN-14-CD is more potent than DHQ in preventing fibrinogen oxidation (in the presence of DHQ absorption peak of fibrinogen is decreased by 12% after ozone oxidation).

For convenience in comparison we create a table that shows relative changes in absorption peak of native fibrinogen, oxidized fibrinogen, a mixture of fibrinogen and KN-14-CD (fibrinogen+KN-14-CD), and oxidized mixture of fibrinogen and KN-14-CD. Also we include data on DHQ activity Table 1.

TABLE 1   The absorption peaks of the studied substances.

Studied compounds	Absorbance at absorption peak	Absorbance, %
Unoxidized fibrinogen	0.65 ± 0.04	100
Oxidized Fibrinogen	0.46 ± 0.03	70
Unoxidized fibrinogen+KN-14-CD	0.66 ± 0.05	100
Oxidized fibrinogen+KN-14-CD	0.63 ± 0.07	95
Unoxidized fibrinogen+DHQ	0.64±0.02	100
Oxidized fibrinogen+DHQ	0.56±0.03	88

Note. Data are presented as mean values and standard errors of the mean (M ± s.e.m.).

To assess the functional activity of fibrinogen after ozone oxidation in the absence and in the presence of KN-14-CD and DHQ, we have studied the ability of fibrinogen to form a fibrin clot after addition of thrombin. The results of this study are presented in Table 2.

TABLE 2   The time of a fibrin clot formation after addition of thrombin to the unoxidized fibrinogen, oxidized fibrinogen, and oxidized fibrinogen mixtures (fibrinogen + KN-14-CD and fibrinogen + DHQ).

T, Un-oxidized fibrinogen, s	T, Oxidized fibrino-gen, s	T, Oxidized fibrinogen+KN-14-CD (5×10⁻⁵ M), s	T, Oxidized fibrinogen+DHQ (5×10⁻⁵ M), s
60 ± 1	–	112 ± 9	115 ± 12

Note. Data are presented as mean values and standard errors of the mean (M ± s.e.m.).

Table 2 shows that if fibrinogen oxidation occurs in the absence of KN-14-CD or dihydroquercetin, then fibrinogen completely loses its functional activity. Whereas if fibrinogen oxidation occurs in the presence of KN-14-CD or dihydroquercetin, then fibrinogen retains its functional properties. These results (DHQ activity) are almost coincident with results of our previous work [18]. There was a less time of clot formation of unoxidized fibrinogen. The fact of different time of an unoxidized fibrin clot formation (as other differences) takes place due to moderate difference in fibrinogen properties.

10.4   CONCLUSION

- In the presence of the new water-soluble dihydroquercetin derivative KN-14-CD ozone oxidation of fibrinogen is negligible (absorption peak of fibrinogen is decreased by 5% in the presence of KN-14-CD, whereas in the case of oxidation without KN-14-CD absorption peak of fibrinogen is decreased by 30%). KN-14-CD is more potent in preventing of fibrinogen oxidation than dihydroquercetin.
- In the presence of KN-14-CD as well as in the presence of dihydroquercetin fibrinogen retains its functional properties after ozone oxidation, whereas in the case of ozone oxidation of fibrinogen without these protective compounds, the functional activity of fibrinogen is completely lost.

KEYWORDS

- **Dihydroquercetin**
- **Fibrinogen**
- **Flavonoids**
- **Hemorheological parameters**

REFERENCES

1. Shacter, E, Williams, J. A.; Levine, R. F. Oxidative modification of fibrinogen inhibits thrombin-catalyzed clot formation. *Fee Radic. Biol. Med.* **1995**, *18*, 815–831.
2. Belisario, M. A.; Di Domenico, C.; Pelagalli, A.; Della Morte, R.; Staiano, N. Metal-ion catalyzed oxidation affects fibrinogen activity on platelet aggregation and adhesion *Biochimie*, **1997**, *79*, 449–455.
3. Roitman E. V.; Azizova O. A.; Morozov Yu. A.; Aseichev A. V. Effect of oxidized fibrinogens on blood coagulation. Bul. of exp. biology and medicine. .**2004**, *138*, 527–530.
4. Rosenfeld M. A.; Leonova V. B.; Shchegolikhin A. N.; Razumovskii S. D.; Konstantinova M. L.; Bychkova A. V.; Kovarskii A. L. Oxidized modification of fragments D and E from fibrinogen induced by ozone. Biochemistry (Moscow), **2010**, *75*, 1285–1293.
5. Azizova, O. A.; Maksyanina, E. V.; Romanov, Y. A.; Aseichev, A. V.; Scheglovitova, O. N. Fibrinogen and its oxidized form induce interleukin-8 [correction of interleukin-2] production in cultured endothelial cells of human vessels. Bul. of exp. biology and medicine. **2004**, *137*, 406–409.
6. Shcheglovitova, O. N.; Azizova, O. A.; Romanov, Y. A.; Aseichev, A. V.; Litvina, M. M.; Polosukhina, E. R.; Mironchenkova, E. V. Oxidized forms of fibrinogen induce expression of cell adhesion molecules by cultured endothelial cells from human blood vessels. *Bul. Exp. Biol. Med.* **2006**, *142*, 277–281.
7. Katie, M.; Weigandt, Nathan White; Dominic Chung; Erica Ellingson; Yi Wang; Xiaoyun Fu; Danilo, C. Pozzo. Fibrin Clot Structure and Mechanics Associated with Specific Oxidation of Methionine Residues in Fibrinogen. *Biophysical J.* **2012**, *103*, 2399–2407.
8. Amalesh Samanta, Gouranga Das, Sanjoy Kumar Das. Roles of flavonoids in plants. *Int J Pharm Sci* Tech, **2011**, *6*, 12–35.
9. Rogovsky, V. S.; Matyushin, A. I.; Shimanovsky, N. L. The prospects of quercetinum and its derivatives administration for prevention and treatment of atherosclerosis. *IMJ*, **2011**, *3*, 114–118.
10. Martin Del Valle E. M. Cyclodextrins and their uses: a review. *Process Biochemistry* **2004**, *39*, 1033–1046.

11. Lushchak V. I. Free radical oxidation of proteins and its relationship with functional state of organisms. Biochemistry (Moscow). **2007,** *72*, 995–1017.

12. Ivanova, I. P.; Knyazev, D. I.; Kudryavtseva, Y. V.; Chuprov, A. D.; Trofimova, S. V. Oxidation of lipids and proteins in lens and blood plasma of rats in ageing. Modern Technologies in Medicine. **2011,** *3*, 16–20.

13. Pandey, K. B.; Mishra, N.; Rizvi, S. I. Protein oxidation biomarkers in plasma of type 2 diabetic patients. Clinical Biochemistry. **2010,** *43*, (4–5), 508–511, March.

14. Stadtman, E. R. Protein oxidation and aging Free Radic. Res. **2006,** *40*, 1250–1258.

15. Rosenfeld, M. A.; Leonova, V. B.; Konstantinova, M. L.; Razumovskii, S. D. Mechanism of enzymatic cross-linking of fibrinogen molecules. Izv. Ros. Akad Nauk, Biol. Ser.; **2008,** *6*, 578–584.

16. Dijkgraaf, L. C.; Zaardeneta, G, Corddewener, F. W.; Liems, R. S.; Schmitz, J. P.; de Bont, L. G.; Milan, S. B. Crosslinking of fibrinogen and fibrinonectin by free radicals: a possible initial step in adhesion formation in osteoarthitis of the temporomandibular joint. *J. Oral Maxillofac Surg.* **2003,** *61*, 101–111.

17. Pretorius, E.; Steyn, H.; Engelbrecht, M.; Swanepoel, A. C.; Oberholzer, H. M. Differences in fibrin fiber diameters in healthy individuals and thromboembolic ischemic stroke patients. *Blood Coagul. Fibrinolysis,* **2011,** *22*(8), 696–700.

18. Koroteev A. M.; Kaziev G. Z.; Koroteev M. P.; Nifant'ev E. E.; Shutov V. M. Russian Patent 2396077, **2009**.

19. Rogovsky, V. S.; Rosenfeld, M. A.; Leonova, V. B.; Konstantinova, M. L.; Razumovsky, S. D.; Matyoushin, A. I.; Shimanovsky, N. L. Dihydroquercetin Inhibits Ozone Oxidation of Fibrinogen, *New Steps in Physical Chemistry, Chem. Phys. and Biochemical Physics,* **2012**.

20. Razumovskii, S. D.; Zaikov, G. E. Ozone and its Reactions with Organic Compounds. Nauka, Moscow, **1974**.

CHAPTER 11

INDUSTRIAL DRYING AND EVAPORATION SYSTEMS

A. K. HAGHI, and G. E. ZAIKOVPETRONIS

CONTENTS

11.1   INTRODUCTION

Drying of wet porous media is coupled in a complicated way. The structure of the solid matrix varies widely in shape. There is, in general, a distribution of void sizes, and the structures may also be locally irregular. Energy transport in such a medium occurs by conduction in all of the phases. Mass transport occurs within voids of the medium. In an unsaturated state these voids are partially filled with a liquid, whereas the rest of the voids contain some gas. It is a common misapprehension that non-hygroscopic fibers (i.e., those of low intrinsic for moisture vapor) will automatically produce a hydrophobic fabric. The major significance of the fine geometry of a textile structure in contributing to resistance to water penetration can be stated in the following manner.

The requirements of a water repellent fabric are: (a) that the fibers shall be spaced uniformly and as far apart as possible and (b) that they should be held so as to prevent their ends drawing together. In the meantime, wetting takes place more readily on surfaces of high fiber density and in a fabric where there are regions of high fiber density such as yarns, the peripheries of the yarns will be the first areas to wet out and when the peripheries are wetted, water can pass unhindered through the fabric.

For thermal analysis of wet fabrics, the liquid is water and the gas is air. Evaporation or condensation occurs at the interface between the water and air so that the air is mixed with water vapor. A flow of the mixture of air and vapor may be caused by external forces, for instance, by an imposed pressure difference. The vapor will also move relative to the gas by diffusion from regions where the partial pressure of the vapor is higher to those where it is lower.

Again, heat transfer by conduction, convection, and radiation and moisture transfer by vapor diffusion are the most important mechanisms in very cool or warm environments from the skin.

Meanwhile, Textile manufacturing involves a crucial energy-intensive drying stage at the end of the process to remove moisture left from dye setting. Determining drying characteristics for textiles, such as temperature levels, transition times, total drying times and evaporation rates, etc., is vitally important so as to optimize the drying stage. In general, drying means to make free or relatively free from a liquid. We define it more narrowly as the vaporization and removal of water from textiles.

11.1.1 HEAT

When a wet fabric is subjected to thermal drying two processes occur simultaneously, namely:
- Transfer of heat to raise the wet fabric temperature and to evaporate the moisture content.
- Transfer of mass in the form of internal moisture to the surface of the fabric and its subsequent evaporation.

The rate at which drying is accomplished is governed by the rate at which these two processes proceed. Heat is a form of energy that can across the boundary of a system. Heat can, therefore, be defined as "the form of energy that is transferred between a system and its surroundings as a result of a temperature difference." There can only be a transfer of energy across the boundary in the form of heat if there is a temperature difference between the system and its surroundings. Conversely, if the system and surroundings are at the same temperature there is no heat transfer across the boundary.

Strictly speaking, the term "heat" is a name given to the particular form of energy crossing the boundary. However, heat is more usually referred to in thermodynamics through the term "heat transfer," which is consistent with the ability of heat to raise or lower the energy within a system.

There are three modes of heat transfer:
- Convection
- Conduction
- Radiation

All three are different. Convection relies on movement of a fluid. Conduction relies on transfer of energy between molecules within a solid or fluid. Radiation is a form of electromagnetic energy transmission and is independent of any substance between the emitter and receiver of such energy. However, all three modes of heat transfer rely on a temperature difference for the transfer of energy to take place.

The greater the temperature difference the more rapidly will the heat be transferred. Conversely, the lower the temperature difference, the slower will be the rate at which heat is transferred. When discussing the modes of heat transfer it is the rate of heat transfer Q that defines the characteristics rather than the quantity of heat.

As it was mentioned earlier, there are three modes of heat transfer, convection, conduction and radiation. Although two, or even all three, modes of heat transfer may be combined in any particular thermodynamic situation, the three are quite different and will be introduced separately.

The coupled heat and liquid moisture transport of nano-porous material has wide industrial applications in textile engineering and functional design of apparel products. Heat transfer mechanisms in nano-porous textiles include conduction by the solid material of fibers, conduction by intervening air, radiation, and convection. Meanwhile, liquid and moisture transfer mechanisms include vapor diffusion in the void space and moisture sorption by the fiber, evaporation, and capillary effects. Water vapor moves through textiles as a result of water vapor concentration differences. Fibers absorb water vapor due to their internal chemical compositions and structures. The flow of liquid moisture through the textiles is caused by fiber-liquid molecular attraction at the surface of fiber materials, which is determined mainly by surface tension and effective capillary pore distribution and pathways. Evaporation and/or condensation take place, depending on the temperature and moisture distributions. The heat transfer process is coupled with the moisture transfer processes with phase changes such as moisture sorption and evaporation.

Mass transfer in the drying of a wet fabric will depend on two mechanisms: movement of moisture within the fabric which will be a function of the internal physical nature of the solid and its moisture content; and the movement of water vapor from the material surface as a result of water vapor from the material surface as a result of external conditions of temperature, air humidity and flow, area of exposed surface and supernatant pressure.

11.1.2 CONVECTION HEAT TRANSFER

A very common method of removing water from textiles is convective drying. Convection is a mode of heat transfer that takes place as a result of motion within a fluid. If the fluid, starts at a constant temperature and the surface is suddenly increased in temperature to above that of the fluid, there will be convective heat transfer from the surface to the fluid as a result of the temperature difference. Under these conditions the temperature difference causing the heat transfer can be defined as:

$$\Delta T = \text{(surface temperature)} - \text{(mean fluid temperature)}$$

Using this definition of the temperature difference, the rate of heat transfer due to convection can be evaluated using Newton's law of cooling:

$$Q = h_c A \Delta T \tag{1}$$

where, A is the heat transfer surface area and h_c is the coefficient of heat transfer from the surface to the fluid, referred to as the "convective heat transfer coefficient."

The units of the convective heat transfer coefficient can be determined from the units of other variables:

$$Q = h_c A \Delta T$$
$$W = (h_c) m^2 K$$

So the units of h_c are $W / m^2 K$.

The relationship given in (Eq.1) is also true for the situation where a surface is being heated due to the fluid having higher temperature than the surface. However, in this case the direction of heat transfer is from the fluid to the surface and the temperature difference will now be

$$\Delta T = \text{(Mean fluid temperature)} - \text{(Surface temperature)}$$

The relative temperatures of the surface and fluid determine the direction of heat transfer and the rate at which heat transfer take place.

As given in Eq. (1), the rate of heat transfer is not only determined by the temperature difference but also by the convective heat transfer coefficient h_c. This is not a constant but varies quite widely depending on the properties of the fluid and the behavior of the flow. The value of h_c must depend on the thermal capacity of the fluid particle considered, i.e., mC_p for the particle. The higher the density and C_p of the fluid the better the convective heat transfer.

Two common heat transfer fluids are air and water, due to their widespread availability. Water is approximately 800 times denser than air and also has a higher value of C_p. If the argument given above is valid then water has a higher thermal capacity than air and should have a better convective heat transfer performance. This is borne out in practice because typical values of convective heat transfer coefficients are as follows:

Fluid	$h_c\left(W\,/\,m^2 K\right)$
water	500–10,000
air	5–100

The variation in the values reflects the variation in the behavior of the flow, particularly the flow velocity, with the higher values of h_c resulting from higher flow velocities over the surface.

11.1.3 CONDUCTION HEAT TRANSFER

If a fluid could be kept stationary there would be no convection-taking place. However, it would still be possible to transfer heat by means of conduction. Conduction depends on the transfer of energy from one molecule to another within the heat transfer medium and, in this sense, thermal conduction is analogous to electrical conduction.

Conduction can occur within both solids and fluids. The rate of heat transfer depends on a physical property of the particular solid of fluid, termed its thermal conductivity k, and the temperature gradient across the medium. The thermal conductivity is defined as the measure of the rate of heat transfer across a unit width of material, for a unit cross-sectional area and for a unit difference in temperature.

From the definition of thermal conductivity k it can be shown that the rate of heat transfer is given by the relationship:

$$Q = \frac{KA\Delta T}{x} \tag{2}$$

ΔT is the temperature difference $T_1 - T_2$,

defined by the temperature on the either side of the porous surface. The units of thermal conductivity can be determined from the units of the other variables:

$$Q = kA\Delta T / x$$
$$W = (k)m^2 K / m$$

The unit of k is $W / m^2 K / m$.

11.1.4 RADIATION HEAT TRANSFER

The third mode of heat transfer, radiation, does not depend on any medium for its transmission. In fact, it takes place most freely when there is a perfect vacuum between the emitter and the receiver of such energy. This is proved daily by the transfer of energy from the sun to the earth across the intervening space.

Radiation is a form of electromagnetic energy transmission and takes place between all matters providing that it is at a temperature above absolute zero. Infrared radiation form just part of the overall electromagnetic spectrum. Radiation is energy emitted by the electrons vibrating in the molecules at the surface of a body. The amount of energy that can be transferred depends on the absolute temperature of the body and the radiant properties of the surface.

A body that has a surface that will absorb all the radiant energy it receives is an ideal radiator, termed a "black body." Such a body will not only absorb radiation at a maximum level but will also emit radiation at a maximum level. However, in practice, bodies do not have the surface characteristics of a black body and will always absorb, or emit, radiant energy at a lower level than a black body.

It is possible to define how much of the radiant energy will be absorbed, or emitted, by a particular surface by the use of a correction factor, known as the "emissivity" and given the symbol ε. The emissivity of a surface is the measure of the actual amount of radiant energy that can be absorbed, compared to a black body. Similarly, the emissivity defines the radiant energy emitted from a surface compared to a black body. A black body would, therefore, by definition, have an emissivity ε of 1. It should be noted that the value of emissivity is influenced more by the nature of texture of clothes, than its color. The practice of wearing white clothes in preference to dark clothes in order to keep cool on a hot summer's day is not necessarily valid. The amount of radiant energy absorbed is more a function of the texture of the clothes rather than the color.

Since World War II, there have been major developments in the use of microwaves for heating applications. After this time it was realized that microwaves had the potential to provide rapid, energy-efficient heating of materials. These main applications of microwave heating today include food processing, wood drying, plastic and rubber treating as well as curing and preheating of ceramics. Broadly speaking, microwave radiation is the term associated with any electromagnetic radiation in the microwave frequency range of 300 MHz-300 Ghz. Domestic and industrial microwave ovens generally operate at a frequency of 2.45 Ghz corresponding to a wavelength of 12.2 cm. However, not all materials can be heated rapidly by microwaves. Materials may be classified into three groups, i.e., conductors' insulators and absorbers. Materials that absorb microwave radiation are called dielectrics, thus, microwave heating is also referred to as dielectric heating. Dielectrics have two important properties:

- They have very few charge carriers. When an external electric field is applied there is very little change carried through the material matrix.
- The molecules or atoms comprising the dielectric exhibit a dipole movement distance. An example of this is the stereochemistry of

covalent bonds in a water molecule, giving the water molecule a dipole movement. Water is the typical case of non-symmetric molecule. Dipoles may be a natural feature of the dielectric or they may be induced. Distortion of the electron cloud around non-polar molecules or atoms through the presence of an external electric field can induce a temporary dipole movement. This movement generates friction inside the dielectric and the energy is dissipated subsequently as heat.

The interaction of dielectric materials with electromagnetic radiation in the microwave range results in energy absorbance. The ability of a material to absorb energy while in a microwave cavity is related to the loss tangent of the material.

This depends on the relaxation times of the molecules in the material, which, in turn, depends on the nature of the functional groups and the volume of the molecule. Generally, the dielectric properties of a material are related to temperature, moisture content, density and material geometry.

An important characteristic of microwave heating is the phenomenon of "hot spot" formation, whereby regions of very high temperature form due to non-uniform heating. This thermal instability arises because of the non-linear dependence of the electromagnetic and thermal properties of material on temperature. The formation of standing waves within the microwave cavity results in some regions being exposed to higher energy than others.

Microwave energy is extremely efficient in the selective heating of materials as no energy is wasted in "bulk heating" the sample. This is a clear advantage that microwave heating has over conventional methods. Microwave heating processes are currently undergoing investigation for application in a number of fields where the advantages of microwave energy may lead to significant savings in energy consumption, process time and environmental remediation.

Compared with conventional heating techniques, microwave heating has the following additional advantages:
- Higher heating rates;
- No direct contact between the heating source and the heated material;
- Selective heating may be achieved;
- Greater control of the heating or drying process.

11.1.5 COMBINED HEAT TRANSFER COEFFICIENT

For most practical situations, heat transfer relies on two, or even all three modes occurring together. For such situations, it is inconvenient to analyze each mode separately. Therefore, it is useful to derive an overall heat transfer coefficient that will combine the effect of each mode within a general situation. The heat transfer in moist fabrics takes place through three modes, conduction, radiation, and the process of distillation. With a dry fabric, only conduction and radiation are present.

11.1.6 POROSITY AND PORE SIZE DISTRIBUTION IN FABRIC

The amount of porosity, i.e., the volume fraction of voids within the fabric, determines the capacity of a fabric to hold water; the greater the porosity, the more water the fabric can hold. Porosity is obtained by dividing the total volume of water extruded from fabric sample by the volume of the sample:

Porosity = volume of water/volume of fabric = (volume of water per gram sample)(density of sample)

It should be noted that most of water is stored between the yarns rather than within them. In the other words, all the water can be accommodated by the pores within the yarns, and it seems likely that the water is chiefly located there. It should be noted that pores of different sizes are distributed within a fabric (Fig. 1). By a porous medium we mean a material contained a solid matrix with an interconnected void. The interconnectedness of the pores allows the flow of fluid through the fabric. In the simple situation ("single phase flow") the pores is saturated by a single fluid. In "two-phase flow" a liquid and a gas share the pore space. As it is shown clearly in Fig.1, in fabrics the distribution of pores with respect to shape and size is irregular. On the pore scale (the microscopic scale) the flow quantities (velocity, pressure, etc.) will clearly be irregular.

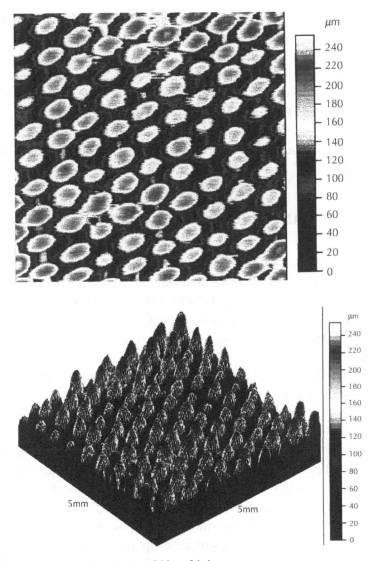

FIGURE 1   Pore size distribution within a fabric.

The usual way of driving the laws governing the macroscopic variables are to begin with standard equations obeyed by the fluid and to obtain the macroscopic equations by averaging over volumes or areas contained many pores.

In defining porosity we may assume that all the pore space is connected. If in fact we have to deal with a fabric in which some of the pore space is disconnected from the reminder, then we have to introduce an "effective porosity," defined as the ratio of the connected pore to total volume.

A further complication arises in forced convection in fabric, which is a porous medium. There may be significant thermal dispersion, i.e., heat transfer due to hydrodynamic mixing of the fluid at the pore scale. In addition to the molecular diffusion of heat, there is mixing due to the nature of the fabric.

11.1.7   THERMAL EQUILIBRIUM AND COMFORT

Some of the issues of clothing comfort that are most readily involve the mechanisms by which clothing materials influence heat and moisture transfer from skin to the environment. Heat flow by conduction, convection, and radiation and moisture transfer by vapor diffusion are the most important mechanisms in very cool or warm environments from the skin.

It has been recognized that the moisture-transport process in clothing under a humidity transient is one of the most important factors influencing the dynamic comfort of a wearer in practical wear situations. However, the moisture transport process is hardly a single process since it is coupled with the heat-transfer process under dynamic conditions. Some materials will posses properties promoting rapid capillary and diffusion movement of moisture to the surface and the controlling factor will be the rate at which surface evaporation can be secured. In the initial stages of drying materials of high moisture content, also it is important to obtain the highest possible rate of surface evaporation. This surface evaporation is essentially the diffusion of vapor from the surface of the fabric to the surrounding atmosphere through a relatively stationary film of air in contact with its surface. This air film, in addition to presenting a resistance to the vapor flow, is itself a heat insulates. The thickness of this film rapidly decreases with increase in the velocity of the air in contact with it whilst never actually disappearing. The inner film of air in contact with the wet fabric remains saturated with vapor so long as the fabric surface has free moisture present. This result in a vapor pressure gradient through the film from the wetted solid surface to the outer air and, with large air movements, the rate of moisture diffusion through the air film will be considerable. The rate of diffusion, and hence evaporation of the moisture will be directly proportional to the exposed area of the fabric, inversely proportional to the film thickness and directly proportional to the inner film surface and the partial pressure of the water vapor in the surrounding air. It is of importance to note at this point that, since the layer of air film in contact with the wetted fabric undergoing drying remains saturated at the temperature of the area of contact, the temperature of the fabric surface whilst still possessing free moisture will lie very close to wet-bulb temperature of the air.

11.1.8 MOISTURE IN FIBERS

The amount of moisture that a fiber can take up varies markedly, as Table 1 shows. At low relative humidifies, below 0.35, water is adsorbed monomolecularly by many natural fibers. From thermodynamic reasoning, we expect the movement of water through a single fiber to occur at a rate that depends on the chemical potential gradient. Meanwhile, moisture has a profound effect on the physical properties of many fibers. Hygroscopic fibers will swell as moisture is absorbed and shrink as it is driven off. Very wet fabrics lose the moisture trapped between the threads first, and only when the threads themselves dry out will shrinkage begin. The change in volume on shrinkage is normally assumed to be linear with moisture content. With hydrophilic materials moisture is found to reduce stiffness and increase creep, probably as a result of plasticization. Variations in moisture content can enhance creep. To describe movement of moisture at equilibrium relative humidity below unity, the idea of absorptive diffusion can be applied. Only those molecules with kinetic energies greater than the activation energy of the moisture-fiber bonds can migrate from one site to another. The driving force for absorptive diffusion is considered to be the spreading pressure, which acts over molecular surfaces in two-dimensional geometry and is similar to the vapor pressure, which acts over three-dimensional spaces.

TABLE 1   Smoothed values of dry-basis moisture content (kg/kg) for the adsorption of water vapor at 30°C onto textile fibers.

Fiber	Mc = 0.2	Mc = 0.5	Mc = 1.0
Cotton	0.0305	0.0565	0.23
Cotton, mercerized	0.042	0.0775	0.335
Nylon 6.6, drawn	0.0127	0.0287	0.05
Orlon (50°C)	0.0031	0.0088	0.05
Cupro	0.0515	0.0935	0.36
Polyester	0.0014	0.0037	0.03
Viscose	0.034	0.062	0.25
Wool	0.062	0.09	0.38

11.2 CONVECTIVE HEAT FLOW

The objective of any drying process is to produce a dried product of desired quality at minimum cost and maximum throughput possible. A very common method of removing water from textiles is convective drying. Hot air is used as the heat transfer medium and is exhausted to remove vaporized water. Considerable thermal energy is required to heat make-up air as the hot air is exhausted. The effect of humidity on the drying rate of textiles depends on both gas stream temperature and flow rate, and are much larger in the warm-up and constant rate periods than in the falling rate period.

11.2.1 INTRODUCTION

When faced with a drying problem on an industrial scale, many factors have to be taken into account in selecting the most suitable type of dryer to install and the problem requires to be analyzed from several standpoints. Even an initial analysis of the possibilities must be backed up by a pilot-scale test unless previous experience has indicated the type most likely to be suitable. The accent today, due to high labor costs, is on continuously operating unit equipment, to what extent possible automatically controlled. In any event, the selection of a suitable dryer should be made in two stages, a preliminary selection based on the general nature of the problem and the textile material to be handled, followed by a final selection based on pilot-scale tests or previous experience combined with economic considerations.

Textile manufacturing involves a crucial energy-intensive drying stage at the end of the process to remove moisture left from dye setting. Determining drying characteristics for textiles, such as temperature levels, transition times, total drying times, and evaporation rates, is vitally important so as to optimize the drying stage. Meanwhile, a textile material undergoes some physical and chemical changes that can affect the final textile quality.

11.2.2 EFFECT OF HUMIDITY ON THE DRYING RATE

A typical drying curve showing moisture content versus time is illustrated in Fig. 1 of Section 11.2. It should be noted that the slope of this curve is the drying rate, at which moisture is being removed. The curve begins with a warm-up period, where the material is heated and the drying rate is usually low. As the material heats up, the rate of drying increases to a peak rate that is maintained for a period of time known as the constant rate period. Eventually, the moisture content of the material drops to a level, known as the critical moisture content, where the high rate of evaporation cannot be maintained. This is the beginning of the falling rate period. During falling rate period, the moisture flow to the surface is insufficient to maintain saturation at the surface. This period can be divided into the first and second falling rate periods. The first falling rate period is a transition between the constant rate period and the second falling rate period. In the constant rate period, external variables such as gas stream humidity, temperature, and flow rate dominate. In the second falling rate period, internal factors such as moisture and energy transport in the textile material dominate.Fig. 1. A typical moisture content profile for textile material.

Although much of the water is removed in the constant rate period of drying, the time required to reduce the moisture in the product to the desired value can depend on the falling rate period. If the target moisture content is significantly lower than the critical moisture content, the drying rates in the falling rate period become important.

The drying process can also be represented by a plot of drying rate versus moisture content, as illustrated if Fig. 2 of Section 11.2. In this plot, time proceeds from right to left. The warm-up period is on the far right, and the constant rate period corresponds to the plateau region. The falling rate period is the section between the plateau region and the origin (Fig. 2 of Section 11.2). A typical drying rate profile for textile material.

11.2.2.1   CONSTANT RATE PERIOD AND FALLING RATE PERIOD

During the constant rate period, evaporation is taking place from the fabric surface. The rate of drying is essentially that of the evaporation of the liquid component under the conditions of temperature and airflow during the process. High air velocities will reduce the thickness of the stationary gas film on the surface of textile material and hence increase the heat and mass transfer coefficients. In commercial forced convection dryers the effects of heat transfer by conduction and radiation may be appreciable, due to the fact that the material surface temperatures are higher than the wet-bulb temperature of the drying air.

As has been stated earlier, the period of constant rate of evaporation from wet textile material is followed by a period during which the rate of drying progressively decreases, the transition from one period to the other taking place at the point of critical moisture content of the textile material. During the constant rate-drying period, surface of the exposed textile material is completely wetted, at the change to falling rate period some of the fabric surfaces will be still wet and some dry depending largely on the physical form of the textile material being dried. The rate of evaporation of the less moist surfaces will be lower than that of the completely wetted portions, the net result being a falling off in the rate of drying as drying proceeds when compared with the rate during the constant rate period. The rate of drying in this part of the drying curve will still be affected by factors which influence the constant rate-drying period as discussed earlier. Once all the exposed surfaces of the textile material cease to be wetted, however, the rate of drying will be a function of the rate at which moisture or moisture vapor can move physically by diffusion and capillarity from within the fabric to its surface.

11.2.3 CONVECTIVE HEAT TRANSFER RATE

Many investigators have attempted to explain the effects of humidity on drying rates and the existence of inversion temperatures. The explanations are usually based on changes that occur in convective heat transfer, radiation heat transfer, and mass transfer as the humidity and temperature of the gas stream change.

At a given gas stream temperature, convective heat transfer rates can change as the humidity in the gas stream is varied, because product temperature and fluid properties vary with humidity. These effects can be explained using the following relationship for the convective heat transfer rate:

$$q / A = h(T_\infty - T_s) = h\Delta T \qquad (1)$$

Here q/A is the convective heat transfer rate per unit surface area A, h is the heat transfer coefficient,

T_∞ Is the free stream temperature of driving medium, T_s is the surface temperature of textile material being dried, and $\Delta T = T_\infty - T_s$ is the temperature difference between the drying medium and the textile material being dried Since product temperature is dependent on humidity, clearly ΔT is also dependent. Further, the heat transfer coefficient h is a function of both product temperature and fluid properties. Thus the convective heat transfer rate changes with humidity, as does the drying rate of a textile material.

11.2.4   EQUILIBRIUM MOISTURE CONTENT

In considering a drying problem, it is important to establish at the earliest stage, the final or residual moisture content of the textile material, which can be accepted. This is important in many hygroscopic materials and if dried below certain moisture content they will absorb or "regain" moisture from the surrounding atmosphere depending upon its moisture and humidity. The material will establish a condition in equilibrium with this atmosphere and the moisture content of the material under this condition is termed the equilibrium moisture content. Equilibrium moisture content is not greatly affected at the lower end of the atmospheric scale but as this temperature increases the equilibrium moisture content figure decreases, which explains why materials can in fact be dried in the presence of super-heated moisture vapor. Meanwhile, drying medium temperatures and humidity assume considerable importance in the operation of direct dryers.

In the more common drying operations met with in practice, the equilibrium moisture content of a material is important as drying may be carried out unnecessarily far, resulting in a reduction in the capacity of a given drying installation and an unjustifiably high cost of drying. Thus, if the equilibrium moisture content of wool hanks in contact with normal ambient air is of the order of 13 to 14% on the wet-weight basis there would be no point in drying to much lower moisture content. Table.1 gives examples of the equilibrium moisture content in contact with air at different percentage relative humidity at 16°C.

It should be noted that two processes occur simultaneously during the thermal process of drying a wet textile material, namely, heat transfer in order to raise temperature of the wet textile material and to evaporate its moisture content together with mass transfer of moisture to the surface of the textile material and its evaporation from the surface to the surrounding atmosphere which, in convection dryers, is the drying medium. The quantity of air required to remove the moisture as liberated, as distinct from the quantity of air which will release the required amount of heat through a drop in its temperature in the course of drying, however, has to be determined from the known capacity of air to pick up moisture at a given temperature in relation to its initial content of moisture. For most practical purposes, moisture is in the form of water vapor but the same principles apply, with different values and humidity charts, for other volatile components Table 1 of Section 11.2.

TABLE 1 Equilibrium moisture content (percentage on dry-weight biases).

Material	20% r.h.a	30% r.h.a	40% r.h.a	50% r.h.a	60% r.h.a
Wool (worsted)	9.5	13.0	16.5	18.5	21.5
Cotton cloth	3.5	4.5	6.0	7.0	7.5
Egyptian cotton	3.5	4.5	5.5	6.0	7.25
Linen	2.75	3.5	4.5	5.1	6.0

r.h.a = relative humidity of air.

11.2.5 INVERSION TEMPERATURE

Drying air will always have an advantage over drying in steam because ΔT in Eq. (1) of Section 11.2, is larger for drying in air; this is a consequence of T_s being very nearly the wet bulb temperature. The wet bulb temperature is lowest for dry air, increases with increasing humidity, and reaches the saturation temperature of water for a pure steam environment.

Thus ΔT_{AIR} will be larger than ΔT_{STEAM}, but $\Delta T_{AIR} / \Delta T_{STEAM}$ decreases with increasing T_∞. Further, the heat transfer coefficient increases with humidity. Apparently, the net effect of the changes in h and ΔT is the convective heat transfer rate increases faster for steam than for air with increasing temperature. Some authors indicated that this is the reason inversion temperature exist. If the dominant heat transfer mechanism is convection, this explanation is plausible.

The variation of gas stream properties with humidity has been reported to explain the existence of inversion temperatures by several investigators.

For example, scientists explain the existence of inversion temperatures based on the total transferable heat in the drying gas. This depends on the specific heat of the gas, the mass flow rate of the gas, and ΔT between the gas and the drying textile material. Further, researchers use a dimensionless correlation for the convective heat transfer coefficient in parallel flow to explain the existence of inversion temperatures. Large Reynolds numbers and Prandl numbers favor higher drying rates in steam, while the high thermal conductivity of air tends to offset these effects. This line of reasoning may be flawed according to scientists, who claim that convective heat transfer is altered in the presence of liquid evaporation. Thus, the correlations developed for heat transfer in the absence of drying may not be applicable to heat transfer with liquid evaporation.

Scientists also have proposed that in version temperatures are due to the higher radiation heat transfer in steam.

Differences in radiation heat transfer are grounded in the Stefan-Boltzmann law. Radiation heat transfer increases with the forth power of temperature, while convective heat transfer varies linearly with temperature. Thus, the relative importance of radiation heat transfer increases with temperature. Since the emissivity of steam is higher than that of air, radiation heat transfer from steam is greater at high temperatures.

11.2.6 MASS TRANSFER

In a specialized sense, the term "mass transfer" is the transport of a substance that is involved as a component (constituent, species) in a fluid mixture. An example is the transport of salt in saltine water. Moreover, convective mass transfer is analogous to convective heat transfer.

Consider a batch of fluid of volume V and mass m. Let the subscript i refer to the ith component of the mixture. The total mass is equal to the sum of the individual masses m_i so $m = \sum m_i$. Hence if the concentration of component i is defined as $C_i = m_i / V$ then the aggregate density ρ of the mixture must be the sum of all the individual concentrations, $\rho = \sum C_i$. Clearly the unit of concentration is kgm^{-3}. Instead of C_i the alternative notation ρ_i is appropriate if we think of each component spread out over the total volume V.

When chemical reactions are of interest it is convenient to work in terms of an alternative description, one involving the concept of mole. By definition, a mole is the amount of substance that contains as many molecules as there are in 12 grams of carbon 12. That number of entities is 6.022×10^{23} (Avogadro's constant). The molar mass of a substance is the mass of one mole of that substance. Hence, if there are n moles in a mixture of molar mass M and mass m, then $n = m / M$. Similarly the number of moles n_i in a mixture is the mass of that component divided by its molar mass M_i, $n_i = m_i / M_i$. The mass fraction of component i is $\phi_i = m_i / m$ so clearly $\sum \phi_i = 1$. Similarly the mole fraction of component i is $x_i = n_i / n$ and $\sum x_i = 1$. To summarize, we have three alternative ways to deal with composition-a dimensional concept (concentration) and two dimensionless ratios (mass fraction and mole fraction). These quantities are related by $C_i = \rho \phi_i = \rho(M_i / M)x_i$, where the equivalent molar mass (M) of the mixture is given by $M = \sum M_i x_i$. If, for example, the mixture can be modeled as an ideal gas, then its equation of state is $PV = mR_m T$ or $PV = nRT$, where the gas constant of the mixture (R_m) and the universal gas constant (R) are related by $R_m = n/m$, $R = R_m / M$. The partial pressure P_i of component i is the pressure we would measure if component i alone were to fill the mixture volume V at the same temperature T as the mixture. Thus $P_i V = m_i R_m T$ or $P_i V = n_i RT$. Summing these equations over i, we obtain Dalton's law, $P = \sum P_i$, which states that the pressure of a mixture of gases at a specified volume and temperature is equal to the sum of the partial pressures of the components. Note that $P_i / P = x_i$, and so using Equation $C_i = \rho \phi_i = \rho(M_i / M)x_i$ and $M = \sum M_i x_i$ we can relate C_i to P_i.

Mass transfer, as used here, is the transfer of moisture from the wet material to the gas stream. One explanation of the existence of inversion temperatures is based on the theory that the driving potential for mass transfer into an environment of steam is different from the driving potential into air.

The driving potential for mass transfer in convective drying in air is commonly considered to be differences of vapor concentration across a boundary layer. An alternative view is that mass transfer in steam occurs by bulk flow due to a pressure difference. The vapor pressure at the surface where evaporation occurs is thought to be slightly higher than the free stream pressure, and causes bulk flow of vapor into the gas stream. Surface temperatures slightly higher than the saturation temperature have been measured.

Only a slight elevation in surface temperature is required to produce a pressure difference capable of generating a large bulk flow of vapor into the gas stream.

11.2.7 DRY AIR AND SUPERHEATED STEAM

A number of investigators have focused on the two humidity extremes, dry air and superheated steam. Comparing drying in superheated steam with drying in air, six observations, excluding economic considerations, warrant attention. At first, the drying rate during the constant rate period.

At saturation temperature, the drying rate in superheated steam is zero, but as the temperature increases, drying rates in superheated steam increases faster than those in air. At the inversion temperature, drying rates in superheated steam increase faster than those in air. Second, the critical moisture content decreases with increasing humidity of the drying gas. Third, in the falling rate period, drying rates in superheated steam roughly equal those in air, even at temperatures below the inversion temperature. Fourth, the equilibrium moisture content of a material dried in superheated steam at high temperatures (usually as high as the inversion temperature) is often lower than the equilibrium moisture content of a textile material dried in air at the same temperature.

As it was mentioned earlier, equilibrium moisture content is the moisture content of a material when it comes to equilibrium with its environmental conditions, primarily the conditions of humidity and temperature. Fifth, more even moisture distribution in the textile material being dried occurs in superheated steam. When drying in air, heterogeneous surface wetting typically occurs below the critical moisture content. Sixth, material porosity and pliability are sustained better when dried in superheated steam than when dried in air.

11.2.8   HEAT SETTING PROCESS

The main aim of the heat setting process is to ensure that fabrics do not alter their dimensions during use. This is particularly important for uses such as timing and driving belts, where stretching of the belt could cause serious problems. It is important to examine the causes of this loss in stability so that a full understanding can be obtained of the effects that heat and mechanical forces have on the stability of fabrics. All fabrics have constraints place on them by their construction and method of manufacture, but it is the heat-setting mechanism that occurs within the fiber that will ultimately influence fabric dimensions.

The temperature and time of setting must be carefully monitored and controlled to ensure the consistent fabric properties are achieved. During heat setting, the segmental motion of the chain molecules of the amorphous regions of the fiber are generally increased leading to structural relaxation within the fiber structure. During cooling, the temperature is decreased below the fiberglass transition temperature (T_g) and the new fiber structure is established. Because the polymer chain molecules have vibrated and moved into new equilibrium positions at a high temperature in heat setting, subsequent heat treatments at lower temperatures do not cause heat-set fiber to relax and shrink, so that the fabric dimensional stability is high. Presetting of fabric prior to dyeing alters the polymer chain molecular arrangement within the fibers, and hence can alter the rate of dye uptake during dyeing. Process variations (e.g., temperature, time or tension differences) during heat setting may thus give rise to dye-ability variations that become apparent after dyeing. Fabric post-setting after coloration can lead to the diffusion of dyes such as disperse dyes to the fiber surface and sublimation, thermo-migration and blooming problems, all of which can alter the color and markedly decrease the color fastness to washing and rubbing of technical textiles containing polyester fibers.

A mixture of superheated steam and air is necessary for the rapid heat setting process. In the steam/air mixture used as the setting medium, the total heat capacity is increased considerably and thus it is possible to apply much more energy to the fabric within a given time and it is possible to heat-set the synthetic fiber fabrics in a shorter time.

An explanation for this increased energy transfer is that the specific heat of steam/air mixture is almost twice of that of air alone. Also by using steam/air mixture, heat exchange at the exterior of the fiber is increased, but the heat conduction from the surface to the interior of the fiber remains unchanged. Thus warp knitted fabrics and lightweight woven like taffeta and georgette require 3–4 seconds with steam/air mixture, compared with 15 seconds if hot air alone is used. Heavier fabrics, especially woven require 9 seconds with rapid heat setting (with steam/air mixture) and 30 seconds with hot air.

Drawn polyester filaments, which have not been set have good tenacity and elasticity properties, but lack dimensional stability when subjected to the action of heat. Also, twisted or doubled filaments have a tendency to curl, which adversely affects further processing. Planar structures produced from unset continuous filaments or fibers exhibit creasing behavior when in use. It is necessary to impart dimensional stability to polyester fabrics so that the garments made from them retain their shape on being subject to washing and ironing – therefore it is necessary to set the drawn, twisted filament or fabric in this state in order to attain resistance against shrinkage, and dimensional curling and crease resistance. A process called heat setting can achieve this in which they are subjected to the action of heat in the presence or absence of swelling agents, with or without tension. In practice, hot water, saturated steam or dry heat is used.

The heat setting process may be explained as follows:

The linkage between the molecular chains or crystallites which were "frozen in" under tension and the mechanical stresses can be balanced by heat, thereby giving rise to greater freedom to the chains to oscillate. This provides the bonds an opportunity to snap into the sites of least energy. Further, the action of heat induces crystallization and orientation processes.

In summery, only by heat-setting the polyester fibers acquire the dimensional stability, crease-resistance and resilience desired in use. It should be noted that heat-setting is usually indispensable for ensuring satisfactory behavior of the material during other finishing processes and is one of the most important finishing processes employed with materials containing polyester fibers or their mixtures with other fibers.

11.2.9   CONVECTIVE HEAT AND MASS TRANSFER COEFFICIENTS

The convective heat and mass transfer coefficients at the surface of textile materials are important parameters in drying processes; they are functions of velocity and physical properties of the drying medium, and in general, can be expressed in the form of Eqs. (2) and (3) of Section 11.2.

$$Nu = a \, Re^b \, Pr^c \qquad (2)$$

$$Sh = a' Re^{b'} \, Sc^{c'} \qquad (3)$$

where, Nu is the Nusselt number, Re, is the Reynolds number, Pr, is the Prandtl number, S is pore saturation and $a, b, c,$ are constants.

It should be noted that for a fully wetted surface, the areas for heat and mass transfer are virtually the same, so that the surface temperature is close to the wet bulb temperature; for partly wetted surface, the effective area for mass transfer decreases with the surface moisture content.

Scientists proposed a model for convective mass transfer coefficient, which assumed that evaporation take places from discontinuous wet surface consisting of dry and wet patches. The ratio of a wet area to the total surface area decreases with decreasing moisture content. However, it is not clear how the fraction of wet area varies with the surface moisture content. Moreover for hygroscopic textile materials, when the fraction of wet area at the surface approaches zero or the surface moisture content is equal to the maximum sorptive value, the evaporation rate at the fabric surface may not be equal to zero.

11.2.10   CONVECTIVE DRYING OF TEXTILE MATERIAL: SIMPLE CASE

In this section we focus on the equations, which consider all the major internal moisture, transfer mechanisms and the properties of the textile material to be dried, including whether it is hygroscopic or non-hygroscopic.

The convective heat and mass transfer coefficients are assumed to vary with the surface moisture content.

As it was mentioned earlier, drying of textile materials involves simultaneous heat and mass transfer in a multiphase system. The drying textile materials may be classified into hygroscopic and non-hygroscopic. For non-hygroscopic textile materials, pores of different sizes form a complex network of capillary paths. The water inside the pores that contribute to flow is called free water. However, the water inside very fine capillaries is difficult to replace by air. This portion of water is known as the irreducible water content. Here, it is defined as bound water. The voids in textile materials are interconnected and filled with air and certain amount of free water.

When a textile material is exposed to convective surface condition, three main mechanisms of internal moisture transfer is assumed to prevail;
- Capillary flow of free water,
- Movement of bound water, and
- Vapor transfer.

If the initial moisture content of the textile material is high enough, the surface is covered with a continuous layer of free water and evaporation takes place mainly at the surface. Internal moisture transfer is mainly attributable to capillary flow of free water through the pores. Therefore, the drying rate is determined by external conditions only, i.e., the temperature, humidity and flow rate of the convective medium, and a constant drying rate period will be observed. As drying proceeds, the fraction of wet area decreases with decreasing surface moisture content, so that the mass transfer coefficient decreases. In order to predict how the wet area fraction varies with surface moisture content, it is necessary to introduce percolation theory.

According to percolation theory, when water passes through randomly distributed paths in a medium, there exists a percolation threshold, which usually corresponds to critical free water movement content. When the free water content is greater than the critical, the water phase is continuous. For a two-dimensional porous medium, this critical value is about 50% of the saturated free water content and for a three-dimensional porous medium it is around 30%. Regardless of the rate of internal moisture transfer, so long as the free water content at the surface is less than the critical, the surface will form discontinuous wet patches. Thus, the mass transfer coefficient decreases with the surface free water content and the first falling rate period's starts. In the first falling rate period, a new energy balance will be reached at the surface, the 'dry' pitches still contain bound water, and the vapor pressure at the surface is determined by the Clausius-Clapeyron equation. When the surface moisture content reaches its maximum value, no free water exists. The surface temperature will rise rapidly, signaling the start of the second falling rate period, during which receding evaporation front often appears, dividing the system into two regions, the wet region and the sorption regions. Inside the evaporation front, the material is wet, i.e., the voids contain free water and the main mechanism of moisture transfer is capillary flow. Outside the front, no free water exists. All water is in the bound water state and the main mechanisms of moisture transfer are movement of bound water and vapor transfer. Evaporation takes place at the front as well as in the whole sorption region, while vapor flows through the sorption region to the surface. Therefore, based on the definitions of the constant rate, first falling rate and second falling rate periods, the characteristics of most drying processes can be described mathematically.

11.2.10.1   CAPILLARY FLOW OF FREE WATER

In textile material, pores provide capillary paths for free water to flow. The driving force for capillary flow is tension gradient or pressure gradient. The pertinent expression for capillary flow of free water is given by:

$$J_L = -\rho_W \frac{K}{\mu}(\nabla P_g - \nabla P_c - \rho_w g) \tag{4}$$

Here,
J_L $[kgm^{-2}s^{-1}]$ represent the free water flux, ρ_w $[kgm^{-3}]$ is the density of water, K $[m^2]$ refers to the permeability, μ $[kgm^{-1}s^{-1}]$ is the viscosity, P_g $[Nm^{-2}]$ is the gas pressure, P_c $[Nm^{-2}]$ is the capillary pressure $\{(P_g - P_L)[Nm^{-2}]\}$ and P_L $[Nm^{-2}]$ is the liquid moisture or free water pressure.

We can also assume that:
- The textile material is macroscopically homogeneous;
- The flow in the capillary paths is laminar;
- There is no significant temperature gradient;
- The effect of the gas phase pressure and gravity force are negligible.

Thus, Eq. (4) can be simplified as:

$$J_L = \rho_w \frac{K_L}{\mu} \left(\nabla P_g - \nabla P_c - \rho_w g \right) \tag{5}$$

As pointed out by Miller and Miller [19], for homogeneous media and negligible gravity forces, tension is proportional to moisture content. It appears that Krischer and Kast's equation for liquid flow may be valid.

$$J_L = -\rho_0 D_L \nabla U \tag{6}$$

where ρ_0 is the bulk density of dry material, $D_L \ [m^2 s^{-1}]$ is the capillary conductivity and $U[kg kg(solid)^{-1}]$ is the moisture content.

Since the permeability K_L depends on the pore structure of the material and the interaction between water and the textile material, it is difficult to find a theoretical form to relate K_L and the capillary conductivity D_L. However, the velocity of flow in capillaries may be assumed to follow the Hagen–Poiseuille law [21]:

$$v = \frac{r^2}{8\mu\tau} \nabla P_c \tag{7}$$

where v ms^{-1} is the fluid velocity, r [m] is the radius and τ is the tortuosity factor of capillary paths.

The total mass flow rate of free water can then be expressed in terms of the local velocity and corresponding capillary radius r_c, which is defined as the radius of the largest capillary in which free water exist, i.e.,

$$J_L = \frac{1}{A} \int \rho_w \frac{r^2}{8\mu\tau} \nabla P_c dA_c = \int_{r_{min}}^{r_c} \rho_w \frac{r^2}{8\mu\tau} \nabla P_c \gamma(r) dr \tag{8}$$

where $\gamma(r)$ is the pore volume density function, defined as:

$$\gamma(r) = \frac{\rho_0}{\rho_w}\frac{\Delta U}{\Delta r}\bigg|_r = \varepsilon\frac{\Delta S}{\Delta r}\bigg|_r$$

where, ε is the porosity and S is the pore saturation. The relationship between the pore saturation and the pore density function is:

$$S = \frac{\int_{r_{min}}^{r_c} \gamma(r)dr}{\int_{r_{min}}^{r_{max}} \gamma(r)dr} = \frac{1}{\varepsilon}\int_{r_{min}}^{r_c} \gamma(r)dr \tag{9}$$

where r_{max} is the overall largest capillary in the porous material. Comparing Eqs. (5) and (6) with Eq. (8), K_L may be expressed as:

$$K_L = -\frac{K_L}{\mu}\frac{\nabla P_c}{\nabla U} = \frac{2}{r_c^2\gamma(r_c)}\frac{\sigma}{\mu}K_L \tag{10}$$

where σ is the surface tension.

The capillary conductivity D_L may be expressed as:

$$D_L = -\frac{K_L}{\mu}\frac{\rho_w\nabla P_c}{\rho_0\nabla U} = \frac{2}{r_c^2\gamma(r_c)}\frac{\sigma}{\mu}K_L \tag{11}$$

The pore volume density function can be obtained from the relationship between capillary pressure and pore saturation of a porous medium. Solving Eqs. (9) and (10) using the correlation of pore volume density obtained from the experimental results , a relationship between pore saturation and relative permeability can be obtained which is very close to the empirical correlation.

$$K_r = \left(\frac{S - S_{ir}}{1 - S_{ir}} \right)^3 \tag{12}$$

Therefore, Eq. (12) can be used to predict the relative permeability, K_r. For water, σ / μ is a linear function of temperature, which can be expressed as:

$$\frac{\sigma}{\mu} 1.064T - 394.3 (ms^{-1}) \tag{13}$$

where T[K] is the temperature. The value of r in Eq. (11) might be constant or a function of free water content. Consequently, the following form is obtained to predict the capillary conductivity of free water for non-hygroscopic textile materials:

$$D_L = (1.604T - 394.3)\beta K_0 \left(\frac{S - S_{ir}}{1 - S_{ir}} \right)^3 \tag{14}$$

where K_0 $[m^2]$ is the single phase permeability of porous material and similarly, for hygroscopic materials:

$$D_L = (1.604T - 394.3)\beta K_0 \left(\frac{U - U_{ms}}{U_0 - U_{ms}} \right)^3 \tag{15}$$

11.2.10.2   MOVEMENT OF BOUND WATER

Movement of bound water, sometimes known as "liquid moisture transfer near dryness" or "sorption diffusion," has been studied by a number of investigators.

It has been shown that liquid moisture transfer still exists in the sorption region and is a strong function of free water content. Scientists studied gas phase convective transport in the dry region, which contains irreducible water and concluded that there could be a liquid moisture flux in the region. Movement of bound water however cannot be simply defined as a diffusion process, which often creates confusion in the analysis of liquid moisture transfer in drying processes. Moreover, the bound water conductivity measured is strongly influenced by moisture content. Therefore, movement of bound water may rather be due to flow along very fine capillaries or through cellular membranes.

In the sorption region, both movement of bound water and vapor transfer play important roles in moisture transfer. The transport equation for bound water may be expressed as:

$$J_b = -\rho_w \frac{K_b}{\mu} \nabla P = -\rho_w \frac{K_b'}{\mu} \nabla P_v \qquad (16)$$

Since the bound water in the sorption region is in equilibrium with vapor in the gas phase, Eq. (16) may be written as:

$$J_b = -\rho_w \frac{K_b'}{\mu} P_v^* \frac{\partial \psi}{\partial U} \nabla U = -\rho_0 D_b \nabla U \qquad (17)$$

where ψ is the relative humidity, $D_b \, [m^2 s^{-1}]$ is bound water conductivity. Thus the bound water conductivity may be written as:

$$D_b = D_{bo} \left(\frac{U_b - U_{eq}}{U_{ms} - U_{eq}} \right)^3 \exp\left(\frac{E_d}{Rt} \right) \qquad (18)$$

where t [s] is time, E_d is defined as the activation energy of movement of bound water.

11.2.10.3   VAPOR FLOW

During a drying process, water vapor flows through the pores of the textile material by convection and diffusion. The equations of vapor flow and airflow may be written as Eqs. (19) and (20).

$$J_v = \frac{P_v M_w}{RT} \frac{K_g}{\mu_g} \nabla P + \frac{D_v M_w}{RT} \nabla P_v \qquad (19)$$

$$J_a = \frac{(P - P_v) M_a}{RT} \frac{K_g}{\mu_g} \nabla P + \frac{m D_v M_a}{RT} \nabla (P - P_v) \qquad (20)$$

where M is the molecular weight, m is the ratio of air and vapor diffusion coefficients.

11.2.11 MACROSCOPIC EQUATIONS GOVERNING HEAT FLOW IN TEXTILE MATERIAL

The theoretical formulation of heat and mass transfer in porous media is usually obtained by a change in scale. We can pass from a microscopic view where the size of the representative volume is small with regards to pores, to a microscopic view where the size of the representative volume ω is large with regard to the pores. Moreover, the heat and mass transfer equations can be deduced from Whitaker's theory. The macroscopic equations can be obtained by averaging the classical fluid mechanics, diffusion and transfer equations over the averaging volume $\omega [m^3]$. The average of a function f is:

$$\bar{f} = \frac{1}{\omega} \int_\omega f d\omega \qquad (21)$$

and the intrinsic average over a phase i is:

$$\bar{f}^i = \frac{1}{\omega_i} \int_\omega f d\omega \qquad (22)$$

11.2.11.1 GENERALIZED DARCY'S LAW

Darcy's law is extended by using relative permeability. For gaseous phase, since no gravitational effect is noted.

$$\overline{v}_g = -\frac{KK_g}{\mu_g}\frac{\partial}{\partial x}(\overline{P}_g^g)$$ (23)

where $\overline{v}_g$ is the speed of the gaseous phase, K $[m^2]$ the intrinsic permeability, K_g the relative permeability to the gaseous phase, $\overline{P}_g^g$ the average intrinsic pressure of the gaseous mixture, and μ_g the viscosity of the gaseous phase.

For the liquid phase:

$$\overline{v}_1 = -\frac{KK_1}{\mu_1}\frac{\partial}{\partial x}(\overline{P}_g^g - \overline{P}_c^1 + \overline{\rho}_1^1 g)$$ (24)

where $\overline{v}_1$ is the speed of the liquid phase, K_1 the relative permeability to the liquid phase, P_c the capillarity pressure, μ_g μ the viscosity of the liquid, g the gravitational constant, and ρ_1 the density of liquid.

11.2.11.2 MASS CONSERVATION EQUATIONS

For the liquid:

$$\frac{\partial\overline{\rho}_1}{\partial t} + \frac{\partial}{\partial x}(\overline{\rho}_v^g\overline{v}_v) = \dot{m}$$ (25)

where $\dot{m}$ is the evaporated water in units of time and volume.
For the vapor:

$$\frac{\partial\overline{\rho}_v}{\partial t} + \frac{\partial}{\partial x}(\overline{\rho}_v^g\overline{v}_v) = \dot{m}$$ (26)

where

$$\bar{\rho}_v^g \bar{v}_v = \bar{\rho}_v^g \bar{v}_g - \bar{\rho}_g^g D_{eff} \frac{\partial}{\partial x}(\bar{\rho}_v / \bar{\rho}_g)$$ (27)

P and P are the average densities of the water vapor and of the gaseous mixture, D the coefficient of the effective diffusion of vapor in the porous medium.

For the gaseous mixture:

$$\frac{\partial \bar{\rho}_g}{\partial t} + \frac{\partial}{\partial x}(\bar{\rho}_g^g \bar{v}_g) = \dot{m}$$ (28)

11.2.11.3　ENERGY CONSERVATION EQUATION

By assuming all the specific heats as constant and with aid of the mass conservation equations, the energy balance takes a form, which is unusual but efficient in the calculations.

$$\frac{\partial}{\partial t}(\bar{\rho}\bar{C}_p\bar{T}) + \frac{\partial}{\partial x}(\bar{\rho}_1^1 C_{p1}\bar{V}_1\bar{T} + \sum_{j=v,a} \bar{\rho}_j^g C_{pj}\bar{v}_j\bar{T})$$
$$= \frac{\partial}{\partial x}(\lambda_{eff} \frac{\partial \bar{T}}{\partial x}) - \Delta h_{vap}^{\circ}\dot{m}$$ (29)

where, Δh_{vap}° is a constant defined by:

$$\Delta h_{vap}^{\circ} = \Delta h_{vap} + (C_{p1} - C_{pv})\bar{T}$$ (30)

λ_{eff} the effective thermal conductivity of the textile material, Δh_{vap} the enthalpy of vaporization, $\bar{\rho}\bar{C}_p$ the constant pressure heat capacity of the textile material.

$$\bar{\rho}\bar{C}_p = \bar{\rho}C_{ps} + \bar{\rho}_1 C_{p1} + \bar{\rho}_v C_{pv} + \bar{\rho}_a C_{pa}$$ (31)

11.2.11.4   THERMODYNAMIC RELATIONS

The partial pressure of the vapor is equal to its equilibrium pressure.

$$\overline{P}_v^{\,g} = \overline{P}_{veq}(T,S) \tag{32}$$

The gaseous mixture is supported to be an ideal mixture of perfect gases.

$$\overline{P}_j = \overline{\rho}_j R\overline{T} / M_j \;\; ; j = \text{a, v} \tag{33}$$

$$\overline{P}_g = \sum_{j=a,v} \overline{P}_j \;\; ; \overline{\rho}_g = \sum_{j=a,v} \overline{\rho}_j \tag{34}$$

11.2.12   HEAT AND MASS TRANSFER OF TEXTILE FABRICS IN THE STENTER

By modification of the mathematical model developed by scientists, the transient temperature and moisture concentration distribution of a fabric in the stenter can be determined.

$$D\frac{\partial^2 C_A}{\partial x^2} = \frac{\partial C_F}{\partial t} + \varepsilon \frac{\partial C_A}{\partial t} \tag{35}$$

and

$$k\frac{\partial^2 T}{\partial x^2} = \rho C_p \frac{\partial T}{\partial t} - \lambda \frac{\partial C_F}{\partial t} \tag{36}$$

Here, $D\,[m^2/s]$ is the diffusion coefficient, $C_A\,[kg/m^3]$ is the moisture content of air in fabric pores, $C_F\,[kg/m^3]$ is the moisture content

of fibers in a fabric, ε is the porosity, $k[W/mK]$ is the thermal conductivity, $\rho[kg/m^3]$ is the density, $C_p[kJ/kgK]$ is the specific heat, $\lambda[kj/kg]$ and is the latent heat of evaporation.
The boundary conditions for convective heat transfer are:

$$q = h_e(T_e - T) \tag{37}$$

and

$$\dot{m} = h_m(C_e - C_A) \tag{38}$$

where, $q[W/m^2]$ is the convective heat transfer rate, $h_e[W/m^2 K]$ heat transfer coefficient, $T_e[K]$ is the external air temperature, $\dot{m}[kg/m^2 s]$ is the mass transfer rate, $h_m[m/s]$ is the mass transfer coefficient, and $C_e[kg/m^3]$ is the moisture content of external air. The deriving force determining the rate mass transfer inside the fabric is the difference between the relative humidities of the air in the pores and the fibers in the fabric. The rate of moisture exchange may be considered as proportional to the relative humidity difference. Thus the rate equation for mass transfer is:

$$\frac{1}{\rho(1-\varepsilon)}\frac{\partial C_F}{\partial t} = K(y_A - y_F) \tag{39}$$

where y_A is the relative humidity of air in pores of fabric and y_F is the relative humidity of fiber of fabric. Also, the relative humidities of air and fabric for polyester are assumed to be:

$$y_A = \frac{C_A RT}{P_s} \tag{40}$$

and

$$y_F = \frac{C_P}{\rho(1-\varepsilon)} \tag{41}$$

where $R[kJ / kgK]$ is the gas constant and $P_s[kg / m^2]$ is the saturation pressure.

The rate constant K in Eq. (39) is an unknown empirical constant. The transient fabric temperatures can be calculated assuming various values of the rate constant K. The value of the rate constant can be varied from 0.01 to 10. When the rate constant is small; the evaporation rate is so small that the moisture content decreases very slowly. Initially, the surface temperature increases rapidly, but later this rate declines (Fig. 3).

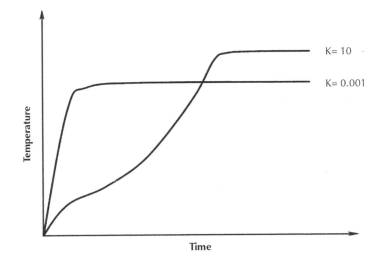

FIGURE 3   Effect of rate constant on the fabric surface temperature.

When K is greater than 1, however, the effect of the rate constant on the surface temperature distribution is not as significant. This indicates that when the rate is greater than 1, the evaporation rate is high and the diffusion mechanism inside the fabric. From (Fig. 4) we see that the surface and center temperatures increase rapidly in the initial stage up to the saturation temperature, at which point the moisture in the fabric starts to evaporate.

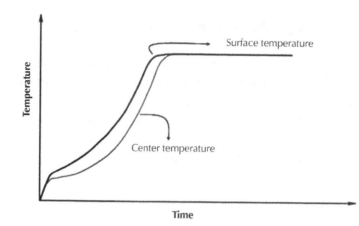

FIGURE 4   Temperature variation of surface and center fabric.

From that point, the difference between the surface temperature and the center temperature increases due to the different moisture contents of the surface and the center. In this stage the fabric stats to dry from the surface and the moisture in the interior is transferred to the fabric surface. Then the moisture content decreases during drying of the fabric. The moisture variations of the surface and the center of fabric are shown in Fig. 5.

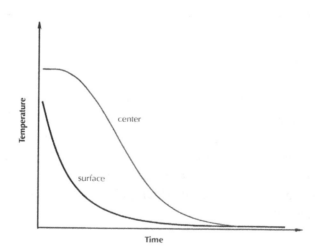

FIGURE 5   Moisture content variation of surface and center fabric.

Initially, the surface moisture content decreases rapidly, but later this rate declines because moisture is transferred to the external air from the fabric surface. The center moisture content remains constant for a short time, and then decreases rapidly, because the moisture content difference between the surface and the interior of the fabric becomes large. After drying out, both center and surface moisture contents converge to reach the external air moisture content. When the initial moisture content is high, the temperature rise is relatively small and drying takes a long time. This may be because the higher moisture content needs much more heat for evaporation from the fabric. Also, the saturation temperature for higher moisture content is lower, and thus the temperature rise in the initial stage is comparatively small (Fig. 6).

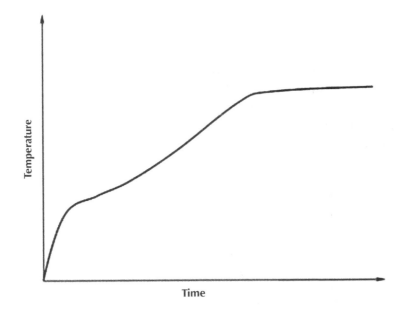

FIGURE 6   Effect of initial moisture content of fabric.

Moreover, the effect of air moisture content can be evaluated as shown in Fig. 7. When the moisture content is high, the initial temperature rise of the fabric also becomes high. This may be because the saturation temperature in the initial stage largely depends on the air moisture content.

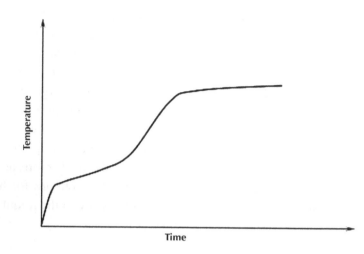

FIGURE 7   Effect of air moisture content.

After the initial temperature rise, however, the temperature increase is relatively small, and thus the time required for complete drying is comparatively long. Meanwhile, when the airflow temperature is high, the temperature rise of the fabric is great (Fig. 8). This may be because the fabric dries more rapidly due to the large difference between its temperature and that of the airflow.

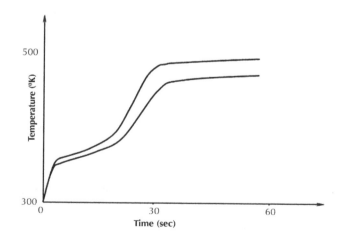

FIGURE 8   Effect of airflow temperature.

11.3   CONDUCTIVE HEAT FLOW

Conduction is the process of heat transfer by molecular motion, supplemented by the flow of heat through textile material from a region of high temperature. Heat transfer by conduction takes place across the interface between two bodies in contact when they are at different temperatures. A common example of heat conduction is heating textile fabric in a cylindrical dryer.

11.3.1   INTRODUCTION

In most textile engineering problems, our primary interest lies not in the molecular behavior of textiles, but rather in how the textile behaves as a continuous medium. In our study of heat conduction, we will therefore neglect the molecular structure of the textile and consider it to be a continuous medium-continuum, which is a valid approach to many practical problems were only macroscopic information is of interest. Such a model may be used provided that the size and the free path of molecules are small compared with other dimensions existing in the medium, so that a statistical average is meaningful. This approach, which is also known as the phenomenological approach to heat conduction, is simpler than microscopic approaches and usually gives the answers required in textile engineering. For heat conduction problems, the use of first and second laws of thermodynamics is sufficient. In addition to these general laws, it is usually necessary to bring certain particular laws into an analysis. There are three such particular laws we employ in the analysis of conduction heat transfer:
- Fourier's law of heat conduction;
- Newton's law of cooling; and
- Stefan-Boltzmann's law of radiation.

11.3.2   FIRST LAW OF THERMODYNAMICS

When a system undergoes a cyclic process, the first law of thermodynamics can be expressed as:

$$\oint \delta Q = \oint \delta W \tag{1}$$

where cyclic integral $\oint \delta Q$ represents the net heat transferred to the system, and the cyclic integral $\oint \delta W$ is the net work done by the system during cyclic process. Both heat and work are path functions. For a process that involves an infinitesimal change of state during a time interval dt, the first law of the thermodynamics is given by:

$$dE = \delta Q - \delta W \tag{2}$$

where δQ and δW are the differential amounts of heat added to the system and the work done by the system, respectively, and dE is the corresponding increase in the total energy of the system during the time interval dt. The energy E is a property of the system and, like other properties, is a point function. That is, dE, depends on the initial and final states only, and not on the path followed between the two states. The physical property E represents the total energy contained within the system and is customarily separated into three parts as bulk kinetic energy, bulk potential energy, and internal energy; that is:

$$E = KE + PE + U \tag{3}$$

     The internal energy U, which includes all forms of energy in a system other than bulk kinetic and potential energies, represents the energy associated with molecular and atomic structure and behavior of the system. Eq. (2) of Section 11.3 can also be written as a rate equation:

$$\frac{dE}{dt} = \frac{\delta Q}{dt} - \frac{\delta W}{dt} \tag{4a}$$

or

$$\frac{dE}{dt} = q - \dot{W} \tag{4b}$$

where $q = \delta Q / dt$ represents the rate of heat transfer to the system and $W = \delta W / dt$ is the rate of work done by the system.

11.3.3   SECOND LAW OF THERMODYNAMICS

The second law leads to the thermodynamic property of entropy. For any reversible process that a system undergoes during a time interval dt, the change in the entropy S of the system is given by:

$$ds = \left(\frac{\delta Q}{T} \right)_{rev} \tag{5a}$$

For an irreversible process, the change, however, is:

$$ds \succ \left(\frac{\delta Q}{T} \right)_{irr} \tag{5b}$$

where δQ is the small amount of heat added to the system during the time interval dt, and T is the temperature of the system at the time of heat transfer. Eq. (5) of Section 11.3 may be taken as the mathematical statement of the second law, and they can also be written in rate form as:

$$\frac{dS}{dt} \geq \frac{1}{T} \frac{\delta Q}{dt} \tag{6}$$

The control-volume from of the second law can be developed also by:

$$\frac{\partial}{\partial t} \int_{cv} s\rho dv + \int_{cs} s\rho V.\hat{n}dA \geq \int_{cs} \frac{1}{T} \frac{\delta Q}{\delta t} \tag{7}$$

where s is the entropy per unit mass, and the equality applies to reversible processes and the inequality to irreversible processes.

11.3.4 HEAT CONDUCTION AND THERMAL CONDUCTIVITY

The rate of heat conduction through a material may be proportional to the temperature difference across the material and to the area perpendicular to the heat flow and inversely proportional to the length of the path of heat flow between the two temperature levels. This dependence was established by well known French scientist J. B. J. Fourier, who used it in his remarkable work, *Theorie Analytique de la Chaleur*, published in Paris in 1822. In this book he gave a very complete exposition of the theory of heat conduction. The constant of proportionality in Fourier's law, denoted by k, is called the thermal conductivity. It is a property of the conducting material and of its state. With the notation indicated in Fig. 1 of Section 11.3, Fourier's law is:

$$q = \frac{kA}{L}(t_1 - t_2)$$

(8)

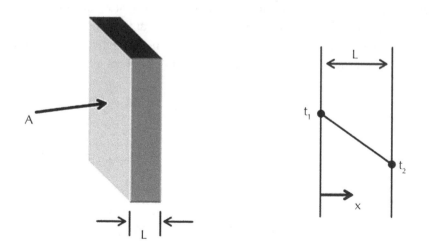

FIGURE 1   One-dimensional steady-state heat conduction.

where kA/L is called the conductance of the geometry. In Fig. 1, since there is a temperature difference of $(t_1 - t_2)$ between the surfaces, heat will flow through the material. From the second law of thermodynamics, we know that the direction of this flow is from the higher temperature surface to the lower one. According to the first law of thermodynamics, under steady conditions, this flow of heat will be at a constant rate.

The thermal conductivity k, which is analogous to electrical conductivity, is a property of the thermal material. It is equivalent to the rate of heat transfer between opposite faces of a unit cube of the material, which are maintained at temperatures differing by 1°. In SI unit, k is expressed as W/mK.

The conduction Eq. (8) may also be written as the heat transfer rate per unit area normal to the direction of heat flow, q," as:

$$\frac{q}{A} = q'' = \frac{k}{L}(t_1 - t_2) = k\left[-\frac{(t_2 - t_1)}{L}\right] = -k\frac{dt}{dx} \qquad (9)$$

The quantity q'' is very useful and is hereafter called the heat flux. The quantity in the brackets is minus the temperature gradient through the material, that is, $-dt/dx$.

Moreover, thermal conductivity is a thermo-physical property. The thermal conductivity of a material depends on its chemical composition, physical structure, and state. It is also varies with the temperature and pressure to which the material is subjected. In most cases, however, thermal conductivity is much less dependent on pressure than on temperature, so that the dependence on pressure may be neglected and thermal conductivity can be tabulated as a function of temperature. In some case, thermal conductivity may also vary with direction of heat flow as in anisotropic materials. The variation of thermal conductivity with temperature may be neglected when the temperature range under consideration is not too severe. For numerous materials, especially within a small temperature range, the variation of thermal conductivity with temperature can be presented by the linear function.

$$k(T) = k_0 \left[1 + \gamma(T - T_0) \right] \tag{10}$$

where $k = k(T_0)$; T_0 is a reference temperature, and γ is a constant called the temperature coefficient of thermal conductivity.

Heat conduction in gases and vapors depends mainly on the molecular transfer of kinetic energy of the molecular movement. That is, heat conduction is transmission of kinetic energy by the more active molecules in high temperature regions to the molecules in low molecular kinetic energy regions by successive collisions. According to kinetic theory of gases, the temperature of an element of gas is proportional to the mean kinetic energy of its constituent molecules. Clearly, the faster the molecules move, the faster they will transfer energy. This implies, therefore, that thermal conductivity of a gas should be dependent on its temperature.

11.3.5 THERMAL CONDUCTION MECHANISMS

Theoretical predictions and measurements have been made of the value of thermal conductivity, k, for many types of substances. In gases, heat is conducted (i.e., thermal energy is diffused) by random motion of molecules. Higher-velocity molecules from higher-temperature regions move about randomly, and some reach regions of lower temperature. By a similar random process, lower-velocity molecules from lower-temperature regions reach higher temperature regions. Thereby, net energy is exchanged between the two regions. The thermal conductivity depends upon the space density of molecules, upon their mean free path, and upon the magnitude of the molecular velocities. The net result of these effects, for gases having very simple molecules, is dependence of k upon T, where T is the absolute temperature. This results from kinetic of gases.

11.3.6 MASS DIFFUSION AND DIFFUSIVITY

Mass diffusion through a region occurs by the motion of quantities of mass or chemical species through the textile material. The diffusion mechanism in a plane layer of textile material is formulated in Fig. 2.

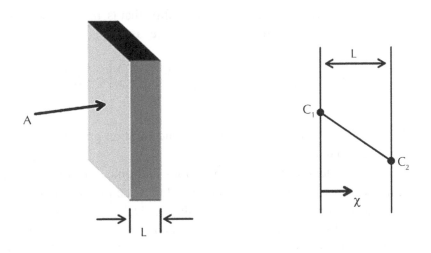

FIGURE 2   Imposed surface concentration C_1 and C_2 at $x = 0$ and L, for a diffusing textile material.

A layer L thick and of area A has concentration levels C_1 and C_2 maintained at the two boundaries, at $x = 0$ and $x = L$. In this example it is assumed that there is no local adsorption or release of stored or bound diffusing material C in the region $x = 0$ to L. The mass diffusion coefficient D is also taken as uniform, that is, constant across the region.

The rate of mass diffusion through the layer is steady state, m, will be proportional to the concentration difference across the material, $(C_1 - C_2)$, and to the area perpendicular to the mass diffusion, A. It will be inversely proportional to the length of the path L of mass diffusion between the two imposed concentration levels C_1 and C_2. Then the mass flow rate through the region, m, of species C is given by:

$$m = \frac{DA}{L}(C_1 - C_2) \tag{10}$$

where DA/L is the "mass" conductance across the region. This result is completely analogous to (Eq.8) for heat flow, which follows from the Fourier Law of Conduction. The mass diffusion formulation given in Eq. (10) of Section 11.3 is a form of Fick's first law.

Eq. (10) is also written in terms of a mass flux, that is, the diffusion rate in mass per unit cross-section area, per unit time, as m":

$$\frac{m}{A} = m" = \frac{D}{L}(C_1 - C_2) = D\left[-\frac{(C_2 - C_1)}{L}\right] = -D\frac{dC}{dx} \quad (11)$$

The quantity in brackets is minus the concentration gradient through the textile material, that is–dc/dx.

Several consistent systems of dimensions and units are used for the physical quantities in Eq. (11) of Section 11.3. A common practice expresses concentration in mass per unit volume, M/L^3, m" in mass flux per unit area and time, that is, M/L^2T. Then D has the dimensions of L^2/T. The unit may be cm^2/s or ft^2/h, or as m^2/s in SI units. Other commonly used measures and terms of composition and concentration include number density of fraction, partial density, mass or mole fraction, and molar concentration.

The magnitude of the diffusion coefficient depends strongly on both the material through which diffusion occurs and the diffusing species. As for heat conduction, as discussed in Eq. (11) indicates that both the gradient of concentration dC/dx and the mass flux $m"(x)$ are constant across the region. This follows from the condition of steady state when D is uniform across the region and no local adsorption of species C occurs in the region.

11.3.7 CONDUCTION HEAT TRANSFER IN TEXTILE FABRIC

In the cylindrical dryers, the transfer of heat to textile fabrics can be calculated by

(Heat transferred to fabric) = (Heat absorbed by fabric) – (Heat transferred from fabric to the ambient air by convection and radiation)

That means,

$$\alpha(T_\infty - T)(dA)(dt) = (m_s)(C)(dA)(dt) + \alpha_t(T - T_a)(dA)(dt)$$

or

$$\alpha(T_\infty - T)dt = (m_s)(C)(dT) + \alpha_t(T - T_a)dt \quad (12)$$

where, $\alpha[Wm^{-2}K^{-1}]$ is the overall heat transfer coefficient of cylindrical dryer-fabric , T[K} is the relative temperature of fabric, T_∞ [K] is the temperature of cylindrical dryer, dA is the differential surface of fabric, t[s] is the heating period, $m_s[kg.m^{-2}]$ is the mass of fabric , $T_a[K]$ is the temperature of ambient and $C[J.kg^{-1}K^{-1}]$ is the specific heat of textile fabric.

It should be noted that thermal loss is an important parameter, which may occur during high rotation of the cylinders. Meanwhile, α_t is a heat transfer coefficient, which characterizes the thermal loss of fabric toward the ambient. This can be shown in the general form of

$$\alpha_t = a(T - T_a) + b \tag{13}$$

Substitution of Eq. (13) of Section 11.3 into Eq. (12) of Section 11.3 yields

$$\alpha(T_\infty - T)dt = (m_s)(C)dT + (a(T - T_a) + b)(T - T_a)dt \tag{14}$$

We assume that α is independent of temperature, then the Eq. (14) of Section 11.3 can be written as:

$$T^2 + K_1T + K_2 \frac{dT}{dt} = K_3 \tag{15}$$

with Eqs. (15)–(20) of Section 11.3.

$$K_1 = \frac{(\alpha + b - 2aT_a)}{a} , \quad K_2 = \frac{(m_s)(C)}{a} , \text{ and}$$

$$K_3 = \frac{\alpha T_\infty + bT_a - aT_a^2}{a}$$

meanwhile, $T_1 = \theta$ could be a particular solution which corresponds to equilibrium temperature $\left(\frac{dT_1}{dt}\right) = 0$.

Therefore,

$$\theta^2 + K_1\theta - K_3 = 0$$

$$\theta_1 = \frac{-K_1 + \sqrt{\delta'}}{2} \tag{16}$$

$$\theta_2 = \frac{-K_1 - \sqrt{\delta'}}{2} \tag{17}$$

with

$$\delta' = K_1^2 + 4K_3 \tag{18}$$

and

$$T = \theta + \frac{1}{(\beta)\exp\left(K_2^{-1}(K_1 + 2\theta)t\right) - (K_1 + 2\theta)^{-1}} \tag{19}$$

To identify θ, we can show,

$$\lim_{t \to \infty} T = \theta$$

θ is equal to the equilibrium temperature of fabric, i.e., $\theta = T_{eq}$.
and

$$\beta = \left(T_a - T_{eq}\right)^{-1} + \left(K_1 + 2T_{eq}\right)^{-1} \tag{20}$$

11.4   RADIATIVE HEAT FLOW

Radiation is a form of electromagnetic energy transmission and takes place between all matters providing that it is at a temperature above absolute zero. Infrared radiation form just part of the overall electromagnetic spectrum. Radiation is energy emitted by the electrons vibrating in the molecules at the surface of a body. The amount of energy that can be transferred depends on the absolute temperature of the body and the radiant properties of the surface.

Electromagnetic radiation is a form of energy that propagates through a vacuum in the absence of any moving material. We observe electromagnetic radiation as light and use it as radio waves, X-rays, etc. Here, we are mostly interested in a form of electromagnetic radiation called microwaves that can be used to heat and dry textile materials.

11.4.1   INTRODUCTION

The word *microwave* is not new to every walk of life as there are more than 60 million microwave ovens in the households all over the world. On account of its great success in processing food, people believe that the microwave technology can also be wisely employed to process materials. Microwave characteristics that are not available in conventional processing of materials consist of: penetrating radiation, controllable electric field distribution, rapid heating, selective heating materials and self-limiting reactions. Single or in combination, these characteristics lead to benefits and opportunities that are not available in conventional processing methods.

Since World War II, there have been major developments in the use of microwaves for heating applications. After this time it was realized that microwaves had the potential to provide rapid, energy-efficient heating of materials. These main applications of microwave heating today include food processing, wood drying, plastic and rubber treating as well as curing and preheating of ceramics. Broadly speaking, microwave radiation is the term associated with any electromagnetic radiation in the microwave frequency range of 300 MHz-300 Ghz. Domestic and industrial microwave ovens generally operate at a frequency of 2.45 Ghz corresponding to a wavelength of 12.2 cm. However, not all materials can be heated rapidly by microwaves. Materials may be classified into three groups, i.e., conductor insulators and absorbers. Materials that absorb microwave radiation are called dielectrics, thus, microwave heating is also referred to as dielectric heating. Dielectrics have two important properties:

- They have very few charge carriers. When an external electric field is applied there is very little change carried through the material matrix.
- The molecules or atoms comprising the dielectric exhibit a dipole movement distance. An example of this is the stereochemistry of covalent bonds in a water molecule, giving the water molecule a dipole movement. Water is the typical case of non-symmetric molecule. Dipoles may be a natural feature of the dielectric or they may be induced. Distortion of the electron cloud around non-polar molecules or atoms through the presence of an external electric field can induce a temporary dipole movement. This movement generates friction inside the dielectric and the energy is dissipated subsequently as heat.

The interaction of dielectric materials with electromagnetic radiation in the microwave range results in energy absorbance. The ability of a material to absorb energy while in a microwave cavity is related to the loss tangent of the material.

This depends on the relaxation times of the molecules in the material, which, in turn, depends on the nature of the functional groups and the volume of the molecule. Generally, the dielectric properties of a material are related to temperature, moisture content, density and material geometry.

An important characteristic of microwave heating is the phenomenon of "hot spot" formation, whereby regions of very high temperature form due to non-uniform heating. This thermal instability arises because of the non-linear dependence of the electromagnetic and thermal properties of material on temperature. The formation of standing waves within the microwave cavity results in some regions being exposed to higher energy than others. Cavity design is an important factor in the control, or the utilization of this "hot spots" phenomenon.

Microwave energy is extremely efficient in the selective heating of materials as no energy is wasted in "bulk heating" the sample. This is a clear advantage that microwave heating has over conventional methods. Microwave heating processes are currently undergoing investigation for application in a number of fields where the advantages of microwave energy may lead to significant savings in energy consumption, process time and environmental remediation.

The benefit of microwave technology has been realized over the past decade with the growing acceptance of microwave ovens in the home. This, together with the gloomy outlook of worldwide energy crises, has paved the way for extensive research into new and innovative heating and drying processes. The use of microwave drying cannot only greatly enhance the drying rates of textile materials, but it may also enhance the final product quality.

While cost presents a major barrier to wider use of microwave in textile industry, an equally important barrier is the lack of understanding of how microwaves interact with materials during heating and drying. The design of suitable process equipment is further confounded by the constraint that geometry places on the prediction of field patterns and hence heating rates within the materials. Effects such as resonance within the material can occur as well as large variations in field patterns at the textile material surface.

The phenomenon of drying has been investigated at considerable length and treated in various texts. However in general, there is only a very small section of this literature devoted to microwave to microwave drying of textile materials.

One of the main features which distinguish microwave drying from conventional drying processes is that because liquids such as water absorb the bulk of the electromagnetic energy at microwave frequencies, the energy is transmitted directly to the wet material. The process does not rely on conduction of heat from the surface of the textile material and thus increased heat transfer occurs, speeding up the drying process. This has the advantage of eliminating case hardening of textile material, which is usually associated with convective hot air drying operations. Another feature is the large increase in the dielectric loss factor with moisture content. This can be used with great effect to produce a moisture leveling phenomenon during the drying process since the electromagnetic energy will selectively or preferentially dry the wettest regions of the solid.

Meanwhile, infrared heating on textile lines has been in use for many years on dyeing lines to pre-dry a host of fabric finishes or topical coatings on fabrics. The renewed interest in infrared pre-drying is due in large part to the need for ever-increasing line speeds and the availability of improved infrared hardware. Infrared predrying of the dyed or finished fabric rapidly preheats and pre-dries wetted fabrics far faster than the typical convection dryer. Typically an air dryer requires 20–25% of its length just to preheat the wetted fabric to a temperature where water is freely evaporated. The infrared pre-heater/pre-dryer section takes over this function in a fraction of the length required in the convection dryer. For dyed fabrics, infrared pre-dryers are typically vertical in configuration, and are generally mounted on the line prior to the drying frame. The systems consist of arrays of electric infrared emitters positioned on both sides of the fabric. The emitters are typically controlled from the fabric temperature. The evaporative load on the pre-dryer dictates how much energy is required and how many vertical sections the pre-dryer must be. With today's more efficient and higher powered emitters most pre-dryers are one or two passes.

It should be noted that controlling shade variations and shade shifts in dyed fabrics has typically been problematic for manufacturing engineers. Dyestuffs tend to migrate to the heated side of the fabric as it passes through the oven. The migration is due partly to gravity, and partly to fluid dynamics. Dyed fabrics come onto the dryer frame at usually 50% to 80% wet pickup. Optimum product quality requires that wet pickup be reduced to the 30% to 60% range with equal water removal from both sides of the fabric. The pre-dried fabric is then presented to the horizontal drying oven with the dyes "locked in" to position. Additional quality benefits can be realized on topical finishes or coatings. Rapid heating with infrared immediately after coating applications tends to keep the coating from deeply wicking into the fabric.

11.4.2   BACKGROUND

This section reviews the basic principles of physics pertaining to microwave heating.

- *Energy*: Energy is the capacitance to do work, and work is defined as the product of force acting over a distance, that is, Eq. (1) of Section 11.4.

$$E = W = (F)(x) \tag{1}$$

Where E = energy, W = the equivalent work, F = force that performs the work, x = distance a mass is moved by the force.

- *Atomic Particles*: All matter are composed of atoms. Atoms, in turn, consist of a nuclei surrounded by orbiting electrons. The nucleus consists of positively charged protons and unchanged neutrons. The surrounding electrons are negatively charged. In neutral atoms, the number of protons in the nucleus equals the number of electrons, resulting in a 0 net charge.
- *Electrostatic Forces*: If some electrons are removed from a piece of material, the protons will outnumber the electrons and the material will take on a positive charge. Similarly, if some electrons are added to a piece of material, the material will take a negative charge. If two positively charged objects are brought near to each other, they will each feel a force pushing them apart. Similarly, if two negatively charged objects are brought together, they will each experience a force pushing them apart. On the other hand, if a negatively charged object is brought near a positively charged object, each will experience a force pulling them together.

- *Columb's law*: If two charges of magnitude q_1 and q_2 are separated by a distance r as shown if Fig. 1 of Section 11.4, each will feel a force magnitude;

$$F = k\frac{q_1 q_2}{r^2} \tag{2}$$

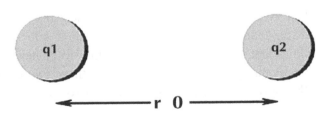

FIGURE 1   Principle of Coulomb's Law.

It is clear from Eq. (2) of Section 11.4 that the force is proportional to the magnitude of each charge and inversely proportional to the square of the distance between them. If, for example, we double the charge on either object, the force will double. On the other hand, if we double the distance between them, the force will be reduced to 1/4 of its previous value.

- *Electric Fields*: Electrostatic force is defined as "force at a distance" Eq.1. If we have a charge Q, and a test charge q is placed a distance R away from it, Q will push on q across that distance as shown in Fig. 2 of section 11.4. The magnitude of push will depend on the magnitudes of Q, q, and r as given in Eq. (2).

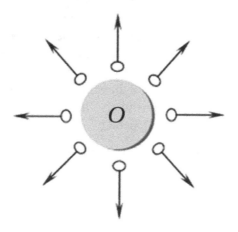

FIGURE 2   Forces around charge Q.

Another way to look at this is to say that Q creates a field in the space that surrounds it. At any point in that space, the field will have a strength E that depends on Q and r. If a test charge is placed at some point in the space, the field at that point will push on it with a force depends on the field strength E at that point and on q. To make these two explanations mathematically equivalent, we separate Eq.2 into two parts; thus,

$$F = k\frac{Qq}{r^2}\left(k\frac{Q}{r^2}\right)(q) \qquad (3)$$

The second part is simply the charge of the second particle. The first part we call E, the field strength at distance r away from Q:

$$E = \left(k\frac{Q}{r^2} \right) \tag{4}$$

Now the force on q can be defined in terms of the field strength times the magnitude of q:

$$F = E.q \tag{5}$$

A microwave oven consists of three major parts:
- The magnetron is the device that generates the microwaves.
- Wave guides direct these waves to the oven cavity.
- The oven cavity holds the material to be heated so that microwaves can impinge on them Eq. (5) of Section 11.4.

Magnetron: It generates microwaves and consists of the following parts:
- *Central cathode*: The cathode is a metal cylinder at the center of the magnetron that is coated with an electron-emitting material. In operation, the cathode is heated to a temperature high enough to cause electrons to boil off the coating.
- *Outer anode*: There is a metal ring called an anode around the magnetron that is maintained at a large positive potential (voltage) relative to the cathode. This sets up an electrostatic field between the cathode and anode that accelerates the electrons toward the anode.

Magnetic field: a strong magnetic field is placed next to the anode and cathode in such an orientation that it produces a magnetic field at right angles to the electrostatic field. This field has the effect of bending the path of the electrons so that, instead of rushing to the anode, they begin to circle in the space between the cathode and anode in a high-energy swarm.

Resonant cavities: they have been built into the anode. Random noise in the electron swarm causes occasional electrons to strike these cavities is such that most radiation frequencies die out. Microwave frequencies, on the other hand, bounce around the cavities and tend to grow, thus getting their energy from the magnetron, passes through the wave guides, and enters the cavity.

However, not all materials can be heated rapidly by microwaves. Materials are reflected from the surface and therefore do not heat metals. Metals in general have high conductivity and are classed as conductors. Conductors are often used as conduits (waveguide) for microwaves. Materials which are transparent to microwaves are classed as insulators. Insulators are often used in microwave ovens to support the material to be heated. Materials which are excellent absorbers of microwave energy are easily used and are classed as dielectric. Fig. 3 of Section 11.4 shows these properties.

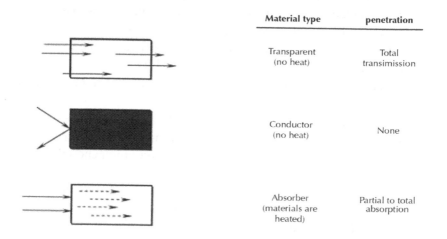

Material type	penetration
Transparent (no heat)	Total transimission
Conductor (no heat)	None
Absorber (materials are heated)	Partial to total absorption

FIGURE 3   Interaction of microwave with materials.

Table 1 of Section 11.4 shows the electromagnetic spectrum. In this continuum, the radio-frequency range is divided into bands as depicted in Table 2 of Section 11.4. Radio-frequency (r.f.) energy has several possible benefits in textile processing. Substitution of conventional heating methods by radio-frequency techniques may result in quicker and more uniform heating, more compact processing machinery requiring less space, and less material in-process at a particular time. Radio-frequency energy has been used for many years to heat bulk materials such as spools of yarn. Bands 9, 10, and 11 constitute the microwave range that is limited on the frequency side by HF and on the high frequency side by the infrared. These microwaves propagate through empty space through empty space at the velocity of light. The frequency ranges from 300 MHz to 300 GHz.

TABLE 1   The electromagnetic spectrum.

Region	Frequencies (Hz)	Wavelength
Audio frequencies	$30 - 30 \times 10^3$	10mm–10km
Radio frequencies	$30 \times 10^3 - 30 \times 10^{11}$	10km–1m
Infrared	$30 \times 10^{11} - 4 \times 10^{14}$	1m–730nm
Visible	$4 \times 10^{14} - 7.5 \times 10^{14}$	730nm–0.3nm
Ultraviolet	$7.5 \times 10^{14} - 1 \times 10^{18}$	400nm–0.3nm
X-rays	$> 1 \times 10^{17}$	< 3nm
Gamma rays	$>1 \times 10^{20}$	< 3nm
Cosmic rays	$>1 \times 10^{20}$	< 3nm

TABLE 2   Frequency bands.

Band	Designation	Frequency limits
4	Very low frequency (VLF)	3–30 kHz
5	Low frequency (LF)	30–300 kHz
6	Medium frequency (MF)	300 kHz– 3MHz
7	High frequency (HF)	3–300 MHz
8	Very high frequency (VHF)	30–300 MHz
9	Ultra high frequency (UHF)	300–3 GHz
10	Super high frequency (SHF)	3–30 GHz
11	Extremely high frequency (EHF)	30–300 GHz

- Pertinent electromagnetic parameters governing the microwave heating:

The loss tangent can be derived from material's complex permittivity. The real component of the permittivity is called the dielectric constant whilst the imaginary component is referred to as the loss factor. The ratio of the loss factor to the dielectric constant is the loss tangent. The complex dielectric constant is given by Eqs. (6) and (7) of Section 11.4:

$$\varepsilon = \varepsilon' - j\varepsilon'' \tag{6}$$

where ε is the complex permittivity, ε' is the real part of dielectric constant; ε'' is the loss factor, and $\varepsilon'/\varepsilon'' = \tan\delta$ is the loss tangent.

Knowledge of a material's dielectric properties enables the prediction of its ability to absorb energy when exposed to microwave radiation. The average power absorbed by a given volume of material when heated dielectrically is given by the equation:

$$P_{av} = \varpi\varepsilon_0\varepsilon_{eff}''E_{rms}^2V \tag{7}$$

Where P_{av} is the average power absorbed (W); ϖ is the angular frequency of the generator (rad/s); ε_0 is the permittivity of free space; ε_{eff}'' is the effective loss factor; E is the electric field strength (V/m); and V is the volume (m^3).

The effective loss factor ε_{eff}'' includes the effects of conductivity in addition to the losses due to polarization. It provides an adequate measure of total loss, since the mechanisms contributing to losses are usually difficult to isolate in most circumstances.

Another important factor in dielectric heating is the depth of penetration of the radiation because an even field distribution in a material is essential for the uniform heating. The properties that most strongly influence the penetration depth are the dielectric properties of the material. These may vary with the free space wavelength and frequency of the propagating wave. For low loss dielectrics such as plastics ($\varepsilon''\ll 1$) the penetration depth is given approximately by:

$$D_P = \frac{\lambda_0\sqrt{\varepsilon'}}{2\pi\varepsilon_{eff}''} \tag{8}$$

Where D_p is the penetration depth; λ_0 is the free space wavelength; ε' is the dielectric constant; and ε''_{eff} is the effective loss factor.

The penetration depth increases linearly with respect to the wavelength, and also increases as the loss factor decreases. Despite this, however, penetration is not influenced significantly when increasing frequencies are used because the loss factor also drops away maintaining a reasonable balance in the above equation

As the material is heated its moisture content will decrease. This leads to a decrease in the loss factor). It can be seen from Eq. (8) of Section 11.4 that the decrease in loss factor causes in the penetration depth of radiation.

Microwaves cause molecular motion by migration of ionic species and/or rotation of dipolar species. Microwave heating a material depends to a great extent on its "dissipation" factor, which is the ratio of dielectric loss or "loss" factor to dielectric constant of the material. The dielectric constant is a measure of the ability of the material to retard microwave energy as it passes through; the loss factor is a measure of the ability of the material to dissipate the energy. In other words "loss" factor represents the amount of input microwave energy that is lost in the material by being dissipated as heat. Therefore, a material with high "loss" factor is easily heated by microwave energy. In fact, ionic conduction and dipolar rotation are the two important mechanisms of the microwave energy loss (i.e., energy dissipation in the material). Non-homogeneous material (in terms of dielectric property) may not heat uniformly, that is, some parts of the materials heat faster than others. This phenomenon is often referred to as thermal runway.

Continuous temperature measurement during microwave irradiation is a major problem. Luxtron fluoroptic or accufiber can be employed to measure temperature up to 400°C but are too fragile for most industrial applications. An optical pyrometer and thermocouple can be employed to measure higher temperatures. Optical pyrometers, such as thermo-vision infrared camera, only records surface temperature, which is invariably much lower than the interior sample temperature. When a thermocouple (metallic probe) is employed for temperature measurements, arcing between the sample and the thermocouple can occur leading to temperature measurements, arcing between the sample and thermocouple can occur leading to failure in thermocouple performance. A recent development is the ultrasonic temperature probe, which covers temperature up to 1500°C.

In summery, microwave heating is unique and offers a number of advantages over conventional heating such as:
- Non-contact heating;
- Energy transfer, not heat transfer;
- Rapid heating;
- Material selective heating;
- Volumetric heating;
- Wuick start-up and stopping;
- Heating starts from interior of the material body.

Some glossaries of microwave heating system are shown in Table 3 of Section 11.4 [4].

TABLE 3   Some glossaries of microwave heating system.

Applicator or cavity	A closed space where a material is exposed to microwaves for heating
Choke	Barriers placed at entrance and exit of the applicator to prevent leakage of microwaves.
Circulator	A three port ferrite device allowing transmission of energy in one direction but directing reflected energy into water load (dummy load) connected at the third port.
Coupling	The transfer of energy from one portion of a circuit to another.
Dielectric	It is a measure of a sample's ability to retard microwave energy as it passes through.
Dielectric loss or loss factor	It is a measure of a sample's ability to dissipate microwave energy.
Hertz (Hz)	1 Hz = 1 cycle/s.
Magnetron	An electronic tube for generating microwaves.
Single mode applicator	Dimension of applicator or cavity is comparable with the wavelength of microwave.
Multimode applicator	An applicator dimension is large in relation to the wave length of incident microwaves.

11.4.3 BASIC CONCEPTS OF MICROWAVE HEATING

As it was mentioned earlier, microwaves are electromagnetic waves having a frequency ranging from 300 MHz and 0.3 THz. Most of the existing apparatuses, however, operate between 400 MHz and 60 GHz, using well-defined frequencies, allocated for industrial, Scientific and Medical (ISM) applications. Among them, the 2.45 GHz is widely used for heating applications, since it is allowed word-wide and it presents some advantages in terms of costs and penetration depth.

It was also mentioned earlier, that quantitative information regarding the microwave-material interaction can be deduced by measuring the dielectric properties of the material, in particular of the real and imaginary part of the relative complex permittivity, $\varepsilon = \varepsilon' - j\varepsilon''_{eff}$, where the term ε''_{eff} includes conduction losses, as well as dielectric losses. The relative permeability is not a constant and strictly depends on frequency and temperature. A different and more practical way to express the degree of interaction between microwaves and materials is given by two parameters; the power penetration depth (D_P) and the power density dissipated in the material (P), as defined earlier in a simplified version as follows:

$$D_P = (\lambda_0 \sqrt{\varepsilon'}) / 2\pi\varepsilon'' \quad P = 2\pi f \varepsilon_0 \varepsilon''_{eff} E_{rms}^2 \tag{9a}$$

where $\varepsilon = \varepsilon' - j\varepsilon''_{eff}$ is the complex permittivity of the material under treatment, λ_0 is the wavelength of the radiation, f is its frequency, $\varepsilon_0 = 8.854.10^{-12}$ F/m is the permittivity of empty space and E_{rms} is the electric field strength inside the material itself. It should be noted that P and D_P can only give quantitative and often misleading information, especially when it is critical to determine the temperature profiles inside the material. Others are the variables involved, however from this two parameters can be deduced most of the peculiarities, which make the microwave, heating a unique process.

First of all, it can be noticed the existence of temperature profile inversion with respect to conventional heating techniques. The air in proximity of the materials during the heat treatment, in fact, is not a good microwave absorber so that it can be considered that the atmosphere surrounding the material is essentially at low temperature.

Vice versa, the material under treatment, interacting in a stronger way with the electromagnetic field, heats up and reaches higher temperature. The result is that, in most cases, the surface temperature of the sample is lower than inside the material itself. This effect is more pronounced for poor heat conducting materials.

Since the given formulation for D_p and P show a strong dependence upon the real and imaginary part of the material permittivity, for multi-phase systems having components with quite different permittivity, it is expected a strong selectivity of the microwave heating process. Power, in fact, is transferred preferentially to some materials (with high $\varepsilon_{eff}^{"}$) so that it can be possible to rise the temperature of just a single phase or component, or to spatially limit the heat treatment to the material, without involving the surrounding environment. This peculiarity can be particularly useful when treating composite materials.

The rapid variations of the permittivity as a function of temperature is responsible for a not always desirable phenomenon, the thermal runaway, that is to say the rapid and uncontrollable overheating of parts of the material under processing. Considering a low thermal conductivity material, whose permittivity increases as the temperature rises, in particular $\varepsilon"$ increasing the temperature growing, it will be subject to gradient of temperature, being colder in the regions where heat is rapidly dissipated or the field strength is lower, and hotter in the remaining zones. These zones, presenting higher values of $\varepsilon"$, and thus of P, will start absorbing microwaves more than the cold ones, further rising their temperature and consequently the local value of $\varepsilon"$, strengthening the phenomenon.

Finally, dielectric heating is penetrating, depending on the operating wavelength, and permits to directly heat treat the surface and the core of the body, without waiting for the heat to reach the core of the sample by means of conduction, particularly time-taking for low thermal conductivity materials, like most polymers are. In these materials, the penetration depth is high, of the order of some tens of centimeters, thus facilitating the processing of large bodies, too.

11.4.4 HEAT AND MASS TRANSFER CLASSICAL EQUATIONS

The conservation of mass and energy for a textile material give the following equations:

$$\frac{\partial X}{\partial t} = D_n \nabla^2 X \tag{9b}$$

$$C_{pn}\rho_n \frac{\partial T}{\partial t} = \nabla(K_n \nabla T) + Q(r,z,t) \tag{10}$$

where $n = 1$ and 2 refers to the inner and outer layer of material, and D is diffusivity (m^2/s); X the moisture content (kg/kg dry basis); k the thermal conductivity ($W/m\,K$); ρ the density (kg/m^3); C_p the heat capacity ($J/kg\,K$); and Q is the microwave source term (W/m^3).

The empirical model for calculating moisture diffusivity as a function of moisture and temperature;

$$D = \frac{1}{1+X} D_0 \exp\left[-\frac{E_0}{R}\left(\frac{1}{T} - \frac{1}{T_r}\right)\right] + \frac{1}{1+X} D_i \exp\left[-\frac{E_i}{R}\left(\frac{1}{T} - \frac{1}{T_r}\right)\right] \tag{11}$$

where D (m^2/s) is the moisture diffusivity; X the moisture content (kg/kg dry basis); T (°C) the material temperature; T_r a reference temperature, and R=0.0083143 kj/mol K is the ideal gas constant; $D_0 (m^2/s)$ the diffusivity at moisture X = 0 and temperature $T = T_r$; D_i (m^2/s) the diffusivity at moisture X=∞ and temperature $T = T_r$; E_0 (kJ/mol) the activation energy of diffusion in dry material at X = 0 and E_i (kj/mol) is the activation energy of diffusion in wet material at X = ∞. The proposed model may uses the estimated parameters in Table 4 of Section 11.4.

TABLE 4   Numerical values for wool (based on data from various authors)

Diffusion coefficient of water vapor — 1st stage:

$(1.04 + 68.20W_c - 1342.59W_c^2)10^{-14}$, $t < 540s$

Diffusion coefficient of water vapor — 2nd stage:

$1.6164\{1 - \exp[-18.163\exp(-28.0W_c)]\}10^{-14}$, $t \geq 540s$

Diffusion coefficient in the air: $2.5e^{-5}$

Volumetric heat capacity of fiber: $373.3 + 4661.0W_c + 4.221T$

Thermal conductivity of fiber:

$(38.49 - 0.720W_c + 0.113W_c^2 - 0.002W_c^3)10^{-3}$

Heat of sorption: $1602.5\exp(-11.72W_c) + 2522.0$

Porosity of fiber: 0.92

Density of fiber: 1300 kg/m^3

Radius of fiber: $1.03 \ e^{-5} m$

Mass transfer coefficient: 0.137 m/s

Heat transfer coefficient: 99.4 $W/m^2 \ K$

Initial conditions: At time t = 0: $T = T_0(r,z)$, $X = X_0(r,z)$

Boundary conditions: $\left.\dfrac{\partial X}{\partial t}\right|_{(r=0, H/2=0, t)} = 0$, $\left.\dfrac{\partial T}{\partial t}\right|_{(r=0, H/2=0, t)} = 0$

Fig. 4 of Section 11.4 shows that there is an increase in drying rate because of the microwave power density.

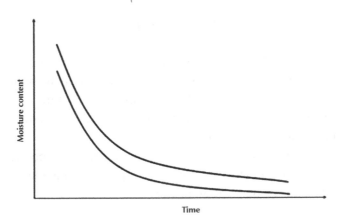

FIGURE 4 Comparison of conventional and microwave heating on average moisture content.

This can be attributed to the effect of microwave on moisture by rapidly increasing the moisture migration to the surface and increased evaporation. A comparison of these drying curves demonstrates improvement in drying times, under microwave heating. Nevertheless the results show significant improvement in average drying times over the conventional heating method.

11.4.5 HEAT AND MASS TRANSFER EXPONENTIAL MODEL

It has been recognized that microwaves could perform a useful function in textile drying in the leveling out of moisture profiles across a wet sample. This is not surprising because water is more reactive than any other material to dielectric heating so that water removal is accelerated. An exponential model presented here can be used to describe the drying curves.

$$X = (a - X_{eq}).\exp(-b.t^d) + X_{eq} \tag{12}$$

and its derivative form:

$$(-dX \,/\, dt) = b.d.t^{(d-1)}.X \qquad\qquad (13)$$

Parameters a, b, d can be determined by regression by the least square method. The quantities b and d vary with the experimental conditions and they are drying coefficients. X is the moisture content of the drying material, dX/dt is the drying rate and t is the drying time. Parameter (a) represents the initial moisture content.

From Fig. 5 of Section 11.4, it is seen that the incident power strongly influenced the drying kinetics of a textile sample, reducing the drying time by raising the microwave heating power.

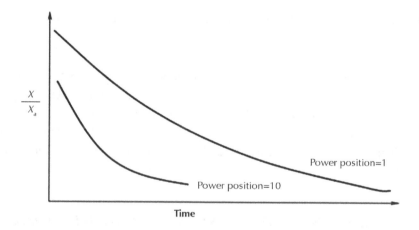

FIGURE 5   Normalized moisture content for two power of microwave heating.

11.4.6 COMBINED MICROWAVE AND CONVECTIVE DRYING OF TUFTED TEXTILE MATERIAL

It should be noted that because of the higher temperature and pressure gradients generated during combined microwave and convective drying, greater care must be taken not to damage the textile material to be dried, whilst still taking advantage of the increased drying rates provided by the microwave environment. To fully understand the heat and mass transfer phenomenon occurring within the material during combined microwave and convective drying, it is required to analyze the moisture, temperature and pressure distributions generated throughout the process

It was shown by scientists that a theory based on a continuum approach led to the following equations of motion governing the drying of a slab of material:

Total mass:

$$\frac{\partial}{\partial t}\left(\phi S_g \rho_g + \phi S_W \rho_W\right) + \nabla.\left(\chi_g \rho_g V_g + \chi_W \rho_W V_W\right) = 0 \quad (14)$$

Total liquid:

$$\frac{\partial}{\partial t}\left(\phi S_g \rho_{gv} + \phi S_W \rho_W\right) + \nabla.\left(\chi_g \rho_{gv} V_{gv} + \chi_W \rho_W V_W\right) = 0 \quad (15)$$

Here, S is the volume saturation, ϕ is the porosity, $\rho[\text{kg}\,m^{-3}]$ is the density of the fibers χ is the surface porosity, ϕ is the porosity

Total enthalpy:

$$\frac{\partial}{\partial t}\left(\phi S_g \rho_{gv} h_{gv} + \phi S_g \rho_{ga} h_{ga} + \phi S_g \rho_{ga} h_{ga} + \phi S_W \rho_W h_W + (1-\phi)\rho_S h_S - \phi \rho_W \int_0^{S_W} \Delta h_W (S) dS\right)$$

$$+ \nabla.\left(\chi_g \rho_{gv} V_{gv} h_{gv} + \chi_g \rho_{ga} V_{ga} h_{ga} + \chi_W \rho_W V_W h_W\right)$$

$$= \nabla.\left((K_g X_g + K_W \chi_W + K_S (1-\chi)\, \nabla T\right) + \phi \quad (16)$$

Where ϕ is the internal microwave power dissipated per unit volume, $K[m^2]$ is permeability, and $h\ [Jkg^{-1}]$ is the averaged enthalpy. In Eq.16 the effects of viscous dissipation and compression work have been omitted.

The Eqs. (14)–(16) of Section 11.4 are augmented with the usual thermodynamic relations and the following relations:

- Flux expressions are given as follows:
Gas flux:

$$\chi_g \rho_g V_g = -\frac{KK_g(S_W)\rho_g}{\mu_g(T)}\left[\nabla P_g - \rho_g g\right] \tag{16a}$$

Here, g [ms^{-2}] is the gravitational constant and K_g is the relative permeability of gas.

Liquid flux:

$$\chi_W \rho_W V_W = -\frac{KK_W(S_W)\rho_W}{\mu_W(T)}\left[\nabla\left(P_g - P_C(S_W,T)\right) - \rho_W g\right] \tag{16b}$$

where, K_W is the relative permeability of water, and μ [Hm^{-1}] is the permeability of free spaces.

Vapor flux:

$$\chi_g \rho_{gv} V_{gv} = \chi_g \rho_{gv} V_{gv} - \frac{\chi_g \rho_g D(T,P_g)M_a M_v}{M^2}\nabla\left(\frac{P_{qv}}{P_g}\right) \tag{16c}$$

Here, V [ms^{-1}] is the averages velocity and M [$kgmol^{-1}$] is the molar mass.

Air flux:

$$\chi_g \rho_{ga} V_{ga} = \chi_g \rho_g V_g - \chi_g \rho_g V_{gv} \tag{16d}$$

- Relative humidity (Kelvin effect) Eqs. (17) and (18) of Section 11.4:

$$\psi(S_W,T) = \frac{P_{gv}}{P_{gvs}(T)} = \exp\left(\frac{2\sigma(T)M_v}{r(S_W)\rho_W RT}\right) \tag{17}$$

where ψ is the relative humidity and $P_{gvs}(T)$ is the saturated vapor pressure given by the Clausius–Clapeyron equation.

- Differential heat of sorption:

$$\Delta h_W = R_v T^2 \frac{\partial (\ln \psi)}{\partial T} \tag{18}$$

- Enthalpy-Temperature relations:

$$h_{ga} = C_{pa}(T - T_R) \tag{19}$$

$$h_{gv} = h_{vap}^0 + C_{pv}(T - T_R) \tag{20}$$

$$h_W = C_{pW}(T - T_R) \tag{21}$$

$$h_s = C_{ps}(T - T_R) \tag{22}$$

The expressions for K_g, K_W are those given by scientists, and μ_g, μ_W have had functional fits according to the data by researchers The diffusivity $D(T, P_g)$ given by scientists and the latent heat of evaporation given by,

$$h_{vap}(T) = h_{gv} - h_W \tag{23}$$

After some mathematical manipulations, the one-dimensional system of three non-linear coupled partial differential equations which model the drying process in a thermal equilibrium environment are given by:

$$a_{s1} \frac{\partial S_W}{\partial t} + a_{s2} \frac{\partial T}{\partial t} = \frac{\partial}{\partial Z}\left[K_{S1} \frac{\partial S_W}{\partial Z} + K_{T1} \frac{\partial T}{\partial Z} + K_{T1} \frac{\partial T}{\partial Z} + K_{P1} \frac{\partial P_g}{\partial Z} + K_{gr1} \right] \tag{24}$$

$$a_{T1} \frac{\partial S_W}{\partial t} + a_{T2} \frac{\partial T}{\partial t} = \frac{\partial}{\partial Z}\left(K_e \frac{\partial T}{\partial Z} \right) - \phi \rho_W h_{vap} \frac{\partial}{\partial Z}\left[K_S \frac{\partial S_W}{\partial Z} + K_T \frac{\partial T}{\partial Z} + K_P \frac{\partial P_g}{\partial Z} K_{gr} \right]$$

$$+\left[\phi\rho_W C_{pW}\left(K_{S2}\frac{\partial S_W}{\partial Z}+K_{T2}\frac{\partial T}{\partial Z}+K_{P2}\frac{\partial P_g}{\partial Z}+K_{gr2}\right)\right]\frac{\partial T}{\partial Z}+\Phi(S_W,T) \quad (25)$$

$$a_{P1}\frac{\partial S_W}{\partial T}+a_{P2}\frac{\partial T}{\partial t}+a_{P3}\frac{\partial P_g}{\partial t}=\frac{\partial}{\partial Z}\left[K_S\frac{\partial S_W}{\partial Z}+K_T\frac{\partial T}{\partial Z}+K_{P3}\frac{\partial P_g}{\partial Z}+K_{gr3}\right] \quad (26)$$

The capacity coefficients a_{S1}, a_{T1}, a_{p1} and the kinetic coefficients K_{S1}, K_{T1}, K_{P1}, K_{gr1} all depend on the independent variables: Saturation S_W, Temperature T and total pressure P_g. The boundary conditions are written in one dimension as:

At z = 0 (Drying surface):

$$K_{S1}\frac{\partial S_W}{\partial Z}+K_{T1}\frac{\partial T}{\partial Z}+K_{P1}\frac{\partial P_g}{\partial Z}+K_{gr1}=\frac{K_m M_V}{R\phi\rho_W}\left(\frac{P_{gV}}{T}-\frac{P_{gV0}}{T_0}\right) \quad (27a)$$

$$K_e\frac{\partial T}{\partial Z}-\phi\rho_W h_{Vap}\left(K_S\frac{\partial S_W}{\partial Z}+K_T\frac{\partial T}{\partial Z}+K_P\frac{\partial P_g}{\partial Z}+K_{gr}\right)=Q(T-T_0) \quad (27b)$$

$$P_g=P_o \quad (27c)$$

At z = L (Impermeable surface):

$$K_{S1}\frac{\partial S_W}{\partial Z}+K_{T1}\frac{\partial T}{\partial Z}+K_{P1}\frac{\partial P_g}{\partial Z}+K_{gr1}=0 \quad (28a)$$

$$K_e\frac{\partial T}{\partial Z}-\phi\rho_W h_{Vap}\left(K_S\frac{\partial S_W}{\partial Z}+K_T\frac{\partial T}{\partial Z}+K_P\frac{\partial P_g}{\partial Z}+K_{gr}\right)=0 \quad (28b)$$

$$\left(K_{S1} - K_S\right)\frac{\partial S_W}{\partial Z} + \left(K_{T1} - K_T\right)\frac{\partial T}{\partial Z} + \left(K_{P1} - K_{P3}\right)\frac{\partial P_g}{\partial Z} + \left(K_{gr1} - K_{gr3}\right) = 0 \quad (28c)$$

Initially:

$$T(z,0) = T_1 \quad\quad (29a)$$

$$P_g(z.0) = P_0 \quad\quad (29b)$$

$$\frac{\partial P_c}{\partial Z} = -\rho_W g \quad\quad (29c)$$

Figs. 6 and 7 of Section 11.4 show compares convective dryings (with or without microwaves). Whilst for convective drying there are definite constant rate and falling rate periods, when microwaves are added the form of the curves change.

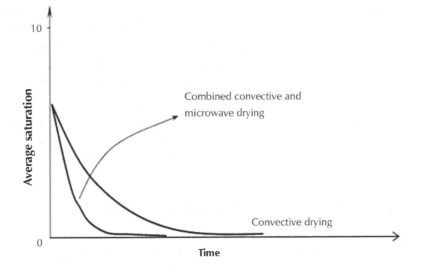

FIGURE 6   Average saturation profiles in time for drying with or without microwaves.

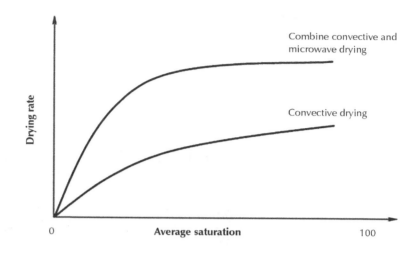

FIGURE 7   Drying rate curves corresponding to profiles plotted in Fig. 6.

11.5   HEAT FLOW AND CLOTHING COMFORT

Information on the transmission of water vapor by textiles fibers is desirable for better understanding of the problems of comfort, and data for design in special applications such as upholstery, footwear, immersion suits and other protective clothing, and wrapping or packaging, where high resistance to liquid water is desired, combined with considerable permeability of water vapor. Some of the issues of clothing comfort that are most readily understood involve the mechanisms by which clothing materials influence heat and moisture transfer from the skin to the environment. Heat transfer by convection, conduction and radiation and moisture transfer by vapor diffusion is the most important mechanisms in very cool or warm environments.

11.5.1   INTRODUCTION AND BACKGROUND

During physical activity the body provides cooling partly by producing insensible perspiration. If the water vapor cannot escape to the surrounding atmosphere the relative humidity of the microclimate inside the clothing increases causing a corresponding increased thermal conductivity of the insulating air, and the clothing becomes uncomfortable. In extreme cases hypothermia can result if the body loses heat more rapidly than it is able to produce it, for example when physical activity has stopped, causing a decrease in core temperature. If perspiration cannot evaporate and liquid sweat (sensible perspiration) is produced, the body is prevented from cooling at the same rate as heat is produced, for example during physical activity, and hyperthermia can result as the body core temperature increases. Table 1 of Section 11.5 shows heat energy produced by various activities and corresponding perspiration rates.

TABLE 1 Heat energy produced by various activities and corresponding perspiration rates [1].

Activity	Work rate (watts)	Perspiration rate (g/day)
Sleeping	60	2280
Sitting	100	3800
Gentle Walking	200	7600
Active Walking	300	11,500
With light pack	400	15,200
With heavy pack	500	19,000
Mountain walking with heavy pack	600–800	22,800–30,400
Maximum work rate	1000–1200	38,000–45,600

The ability of fabric to allow water vapor to penetrate is commonly known as breath-ability. This should more scientifically be referred to as water vapor permeability. Although perspiration rates and water vapor permeability are usually quoted in units of grams per day and grams per square meter per day, respectively, the maximum work rate can only be endured for a very short time. During rest, most surplus body heat is lost by conduction and radiation, whereas during physical activity, the dominant means of losing excess body heat is by evaporation of perspiration. It has been found that the length of time the body can endure arduous work decreases linearly with decrease in fabric water vapor permeability.

It has also been shown that the maximum performance of a subject wearing clothing with a vapor barrier is some 60% less than that of a subject wearing the same clothing but without a vapor barrier. Even with two sets of clothing that exhibit a small variation in water vapor permeability, the differences in the wearer's performance are significant.

In an environment where body temperature cannot be regulated without a lot of sweating, we often try to get rid of heat from our body by turning on the air conditioning systems or moving into a conditioned room. Just after the change of the environment, we will feel "cool" or "comfortable." But the sweat accumulated in clothing evaporates gradually, until the heat loss from our body can be more than needed and at last we might feel "cold" or uncomfortable." A review of clothing studies has shown that moisture collection in cold weather clothing, even after heavy exercise, seldom exceeds 10% by weight of added water. One of the measurements is used to calculate values related to water vapor transmission properties are "water vapor resistance." This is the water vapor pressure difference across the two faces of the fabric divided by the heat flux per unit area, measured in square meters Pascal per watt. Some water vapor resistance data on different types of outwear fabrics are presented in Table 2 of Section 11.5. The measurement of water vapor resistance in thickness unit (mm) is the thickness of a still air layer having the same resistance as the fabric. The use of thickness units facilitates the calculations of resistance values for clothing assemblies comprising textile and air layers.

TABLE 2   Typical water vapor resistance (WVR) of fabrics [1].

Fabric, Outer (shell) material	WVR (mm still air)
Neoprene, rubber or PVC coated	1000–1200
Conventional PVC coated	300–400
Waxed cotton	1000+
Wool overcoating	6–13
Leather	7–8

TABLE 2 *(Continued)*

Woven microfiber	3–5
Closely woven cotton	2–4
Ventile L28	3.5
Other Ventile	1–3
Two-layer PTFE laminates	2–3
Three-layer laminates (PTFE, polyester)	3–6
Microporous polyurethane (various types)	3–14

This observation is generally explained by noting that the major transfer mechanism from wet skin to underwear is one of distillation. An initial observation noting the surprisingly strong discomfort sensations associated with small amounts of water in the skin-clothing interface.

It has been confirmed in a number of studies in which either moisture from sweating or added moisture generates these clothing contact sensations. The procedures for these measurements emphasize again that very little moisture is required to stimulate sensations of discomfort. Often 3% to 5% added moisture is ample to develop discomfort.

Simultaneous differential equations for the transfer of heat and moisture in porous medial under combined influence of gravity and gradients of temperature and moisture content were developed by Researchers. They have performed a general analysis of moisture migration in a slab of an unsaturated porous material for a condition where the temperature of one surface is suddenly increased to a higher value whereas the temperature of the other surface is maintained constant. Researchers derived a general, one-dimensional, steady-state model describing the heat and mass transfer within a homogeneous porous medium, saturated with a wetting liquid, its vapor and a non-condensable gas. The effects of gas diffusion, phase change, conduction, liquid and vapor transport, capillarity, and gravity are included. The analysis is based on a general thermodynamic description of the unique equilibrium states characteristics of liquid wetting porous media. Scientists have provided a systematic, rigorous and unified treatment of the governing equations for simultaneous heat and mass transfer within a wide range of porous media.

Some work has also been done in the area of coupled diffusion of moisture and heat in hygroscopic textile materials. Scientists have given a review of numerical modeling of convection, diffusion and phase changes in textiles. The paper summarizes current and past work aimed at utilizing CFD techniques for clothing applications. It was shown that water in a hygroscopic porous textile might exist in vapor or liquid form in the pore spaces. Phase changes associated with water include liquid evaporation/condensation in the pore spaces and sorption/desorption from polymer fibers. Additional factors such as swelling of solid polymer due to water and heat of sorption was incorporated into the appropriate conservation and transport equations. Scientists have attempted to solve the non-linear differential equations, which describe coupled diffusion of heat and mass (moisture) in hygroscopic textile materials. In addition to the diffusion equations, a rate equation was introduced describing the rate of exchange of moisture between the solid (textile fibers) and the gas phase. The predictions compared favorably with experimental observations on wool bales and wool fabrics. Researchers developed a simple model of combined heat and water vapor transport in clothing. Transport by forced convection was not included in this model.

Scientists reported a strong dependency of water vapor resistance of hydrophilic membranes or coatings: the higher the relative humidity at the membrane, the lower the water vapor resistance (i.e., the higher the water vapor permeability or breathability).

In a temperature dependent experiment, Scientists placed a hydrophilic film on an ice block. Water vapor sublimating from the ice could diffuse only through the film and was collected by a desiccant. Researchers measured mass transport through the film, and he found that water vapor resistance is an exponential function of temperature. In this experiment, water vapor permeability varnishes nearly completely with decreasing textile temperature. Because diffusion in hydrophilic materials is non-Fickian, he also derived from his results a theory of diffusion speed depending on activation energy, and he accounted for different relative humidity. Additionally, scientists reported an increase in the moisture vapor transmission rate of hydrophilic and microporous textiles when measuring with a heated dish instead of unheated dish. They interpreted their results by the increased motion of water vapor and polymer molecules, which they claimed would also work for micro-porous constructions.

Researchers compared cotton, water repellent cotton, and acrylic garments through wearing tests and concluded that the major factor causing discomfort was the excess amount of sweat remaining on the skin surface. Scientists stated that the ability of fabrics to absorb liquid water (sweat) is more important than water vapor permeability in determining the comfort factor of fabrics.

Scientists postulated physiological factors related to the wearing comfort of fabrics as follows: sweating occurs whenever there is a tendency for the body temperature to rise, such as high temperature in the surrounding air and physical exercise, etc. If liquid water (sweat) cannot be dissipated quickly, the humidity of the air in the space in between the skin and the fabric that contacts with the skin rises. This increased humidity prevents rapid evaporation of liquid water on the skin and gives the body the sensation of "heat" that triggered the sweating in the first place. Consequently, the body responds with increased sweating to dissipate excess thermal energy. Thus a fabric's inability to remove liquid water seems to be the major factor causing uncomfortable feelings for the wearer.

Scientists conducted wearer trails for shirts made of various fibers. They concluded that the largest factor that influenced wearing comfort was the ability of fibers to absorb water; regardless of weather fibers were synthetic or natural.

All of these studies indicate that the transient state phenomenon responding to the physiological demand to cause sweating is most relevant to comfort or discomfort associated with fabrics.

When work is performed in heavy clothing, evaporation of sweat from the skin to the environment is limited by layers of wet clothing and air. The magnitude of decrement in evaporative cooling is a function of the clothing's resistance to permeation of water vapor.

Scientists conducted an experimental study on the rate of absorption of water vapor by wool fibers. They observed that, if a textile is immersed in a humid atmosphere, the time required for the fibers to come to equilibrium with this atmosphere is negligible compared with the time required for the dissipation of heat generated or absorbed when the regain changes. Scientists investigated the effects of heat of sorption in the wool-water sorption system. They observed that the equilibrium value of the water content was directly determined by the humidity but that the rate of absorption and adsorption decreased as the heat-transfer efficiency decreased. Heat transfer was influenced by the mass of the sample, the packing density of the fiber assembly, and the geometry of the constituent fibers. Scientists pointed out that the water-vapor-uptake rate of wool is reduced by a rise in temperature that is due to the heat of sorption. The dynamic-water-vapor-sorption behavior of fabrics in the transient state will therefore not be the same as that of single fibers owing to the heat of sorption and the process to dissipate the heat released or absorbed.

Scientists started theoretical investigation of this phenomenon. They proposed a system of differential equations to describe the coupled heat and moisture diffusion into bales of cotton. Two of the equations involve the conservation of mass and energy, and the third relates fiber moisture content with the moisture in the adjacent air. Since these equations are non-linear, they made a number of simplifying assumptions to derive an analytical solution.

In order to model the two-stage sorption process of wool fibers, scientists proposed three empirical expressions for a description of the dynamic relationship between fiber moisture content and the surrounding relative humidity. By incorporating several features omitted by Henry into the three equations, researchers were able to solve the model numerically. Since their sorption mechanisms (i.e., sorption kinetics) of fibers were neglected, the constants in their sorption-rate equations had to be determined by comparing theoretical predictions with experimental results. Based on conservation equations, this global model consists of two differential coupled equations with variables for temperature and water concentration in air (C_a) and in the fibers of the textile (C_f), which is generally the water adsorbed by hygroscopic fibers. C_f is not in equilibrium with C_a, but an empirical relation between the adjustable parameters is assumed: the rate of sorption is a linear function of the difference between the actual C_f and the equilibrium value. The introduced coefficients are not directly linked to the physical properties of the clothes.

Scientists reported a numerical model describing the combined heat and water-vapor transport through clothing. The assumptions in the model did not allow for the complexity of the moisture-sorption isotherm and the sorption kinetics of fibers. Researchers presented two mechanical models to simulate the interaction between moisture sorption by fibers and moisture flux through the void spaces of a fabric. In the first model, diffusion within the fiber was considered to be so rapid that the fiber moisture content was always in equilibrium with the adjacent air. In the second model, the sorption kinetics of the fiber was assumed to follow Fickian diffusion. In these models, the effect of heat of sorption and the complicated sorption behavior of the fibers were neglected.

Scientists developed a two-stage model, which takes into account water-vapor-sorption kinetics of wool fibers and can be used to describe the coupled heat and moisture transfer in wool fabrics. The predictions from the model showed good agreement with experimental observations obtained from a sorption-cell experiment. More recently, Scientists further improved the method of mathematical simulation of the coupled diffusion of the moisture and heat in wool fabric by using a direct numerical solution of the moisture-diffusion equation in the fibers with two sets of variable diffusion coefficients. These research publications were focused on fabrics made from one type of fiber. The features and differences in the physical mechanisms of coupled moisture and heat diffusion into fabrics made from different fibers have not been systematically investigated.

Scientists compared the heat exchange and thermal insulation of two ensembles, one made from wool, the other from nylon, worn by subjects who exercised either lightly (dry condition) or strenuously (wet condition) for 60 minutes, then rested 60 minutes. He found that there was a significant difference in physiological and subjective responses between dry and wet conditions, but not between the two fiber types. Further, there was no significant difference between the ratings of temperature and humidity sensations for the wool and nylon garments. The wool garment picked up more water than the nylon garment (245 g versus 198 g) for the wet condition.

However, the wool fabric may have been slightly thicker than the nylon fabric, since it was reported to have a slightly greater thermal resistance and therefore hold more water.

Scientists evaluated the effect of five kinds of knit structures, all made from 100% polypropylene were evaluated. On subjects exercising for 40 minutes at 5°C followed by 20 minutes at rest, and then repeated. The thickest knit, a fleece, caused the greatest total sweat production, retained the most moisture, and wetted skin the most. They stated that the hydrophobic polypropylene prevented extensive sweat accumulation in the underwear (10 to 22%) causing the sweat to accumulate in the outer garments.

Scientists repeated the protocol above, but used low and high work rates with three kinds of underwear (a polypropylene 1×1 knit, a wool 1×1 knit, and a fishnet polypropylene) worn under wool fleece covered by polyester/cotton outer garments. Total sweat production and evaporated sweat were the same for all three underwear fabrics, but where the sweat accumulated differed significantly. More sweat accumulated in the wool underwear than either polypropylene at both work rates. At the higher work rate, more sweat moved into the fleece layer from both kinds of polypropylene underwear than for the wool. Most likely for the 1×1 knits, the thicker wool underwear (1.95 mm) simply holds more water than the polypropylene underwear (1.41 mm) and based on outer layer-to layer wicking results, needs a greater volume of sweat to fill it pores before it starts to donate the excess to the layer above it.

Scientists conducted wear trails for shirts made of various fibers. They concluded that the largest factor that influenced wearing comfort was the ability of fibers to absorb water, regardless of whether fibers were synthetic or natural.

All of these studies indicate that the transient state phenomenon responding to the physiological demand to cause sweating is most relevant to comfort or discomfort associated with this general principle. It is important to point out that a highly water absorbing fabric placed in the first layer keeps the partial pressure of water vapor near the skin low, which helps dissipate water at the skin surface, although the water vapor transport rate is smaller than for non-absorbing fabrics.

In the other words, the dissipation of water by means of absorption by fabrics appears to be much more efficient way to keep the water vapor pressure near the skin low than dissipation by permeation through fabrics. Highly water-absorbing fabrics raise the temperature of the air space near the skin. The temperature rise will further decrease relative humidity;

however, the higher temperature may or may not desirable depending on environmental conditions.

In the literature, the emphasis has been placed on the correlation between sweating and discomfort associated with wearing fabrics. However, there is relatively less emphasis placed on the influence of changes in the surrounding conditions, that is, the influence of the seasons. Many comfort studies are conducted with a single layer of fabric at relatively warm and moderately humid conditions. Severe winter conditions, which mandate the use of layered fabrics, would necessitate totally different kinds of testing procedures. Consequently, it is necessary to distinguish the comfort factor and the survival factor, and to investigate these factors with different perspective.

The evaporation process is also influenced by the liquid transport process. When liquid water cannot diffuse into the fabric, it can only evaporate at the lower surface of the fabric. As the liquid diffuses into the fabric due to capillary action, evaporation can take place throughout the fabric.

Moreover, the heat transfer process has significant impact on the evaporation process in cotton fabrics but not in polyester fabrics. The process of moisture sorption is largely affected by water vapor diffusion and liquid water diffusion, but not by heat transfer. When there is liquid diffusion in the fabric, the moisture sorption of fibers is mainly determined by the liquid transport process, because the fiber surfaces are covered by liquid water quickly. Meanwhile, the water content distributions in the fibers are not significantly related to temperature distributions.

All moisture transport processes, on the other hand, affect heat transfer significantly. Evaporation and moisture sorption have a direct impact on heat transfer, which in turn is influenced by water vapor diffusion and liquid diffusion. The temperature rise during the transient period is caused by the balance of heat released during fiber moisture sorption and the heat absorbed during the evaporation process.

As a whole, a dry fabric exhibits three stages of transport behavior in responding to external humidity transients. The first stage is dominated by two fast processes: water vapor diffusion and liquid water diffusion in the air filling the inter-fiber void spaces, which can reach new steady states within fractions of seconds. During this period, water vapor diffuses into the fabric due to the concentration gradient across the two surfaces. Meanwhile, liquid water starts to flow out of the regions of higher liquid content to the dryer regions due to surface tension force.

The second stage features the moisture sorption of fibers, which is relatively slow and takes a few minutes to a few hours to complete. In this period, water sorption into the fibers takes place as the water vapor diffuses into the fabric, which increases the relative humidity at the surfaces of fibers. After liquid water diffuses into the fabric, the surfaces of the fibers are saturated due to the film of water on them, which again will enhance the sorption process. During these two transient stages, heat transfer is coupled with the four different forms of liquid transfer due to the heat released or absorbed during sorption/adsorption and evaporation/condensation. Sorption/ adsorption and evaporation/condensation, in turn, are affected by the efficiency of the heat transfer. For instance, sorption and evaporation in thick cotton fabric take a longer time to reach steady states than in thin cotton fabrics.

Finally, the third stage is reached as a steady state, in which all four forms of moisture transport and the heat transfer process become steady, and the coupling effects among them become less significant. The distributions of temperature, water vapor concentration, fiber water content, and liquid volume fraction and evaporation rate become invariant in time. With the evaporation of liquid water at the upper surface of the fabrics, liquid water is drawn from capillaries to the upper surface.

11.5.2   EFFECTIVE THERMAL CONDUCTIVITY

One way of expressing the insulating performance of a textile is to quote "effective thermal conductivity." Here the term "effective" refers to the fact that conductivity is calculated from the rate of heat flow per unit area of the fabric divided by the temperature gradient between opposite faces. It is not true condition, because heat transfer takes place by a combination of conduction through fibers and air and infrared radiation. If moisture is present, other mechanisms may be also involved. Research on the thermal resistance of apparel textiles, has established that the thermal resistance of a dry fabric or one containing very small amounts of water depends on its thickness, and to lesser extent on fabric construction and fiber conductivity. Indeed, measurements of effective thermal conductivity by standard steady-state methods show that differences between fabrics and mainly attributable to thickness. Despite these findings, consumers continue to regard wool as "warmer" than other fibers, and show preference for wearing wool garments in cold weather, particularly when light rain or sea spray is involved.

Meanwhile, the effective thermal conductivities of fabrics can be studied for varying regains. Regain is the mass of water present expressed as a percentage of the dry weight of the material. The effective thermal conductivities for porous acrylic, polypropylene, wool, and cotton is shown in Fig. 1 of Section 11.5.

The curves indicate that changes in "effective thermal conductivity" with increasing regain are not linear, but can be explained in terms of water within the fibers of fabrics with regain.

Figure 2 of Section 11.5 presents the various phases diagrammatically. When fabrics containing water are subjected to a temperature gradient, three different modes of heat flow can be distinguished.

- The presence of condensed water;
- Vapor transport; and
- Condensation.

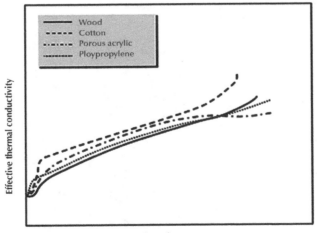

FIGURE 1 Comparison of effective thermal conductivities for porous acrylic, polypropylene, wool and cotton.

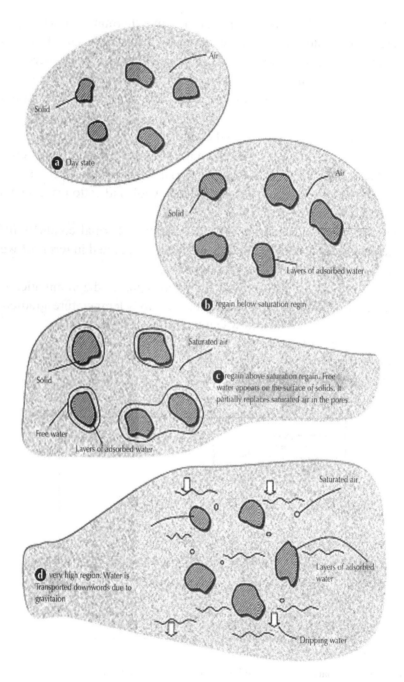

FIGURE 2   *(Continued)*

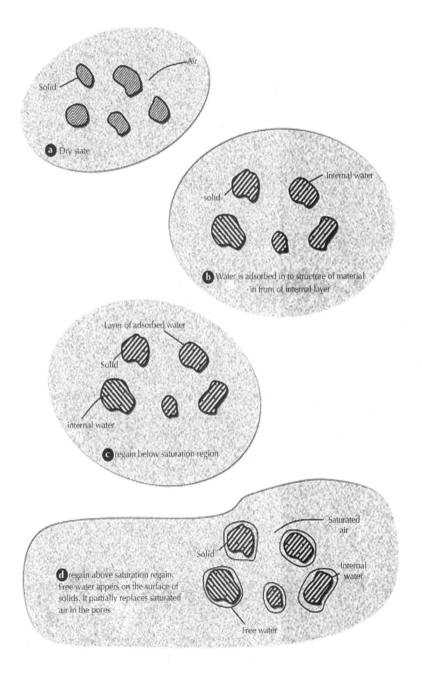

FIGURE 2   *(Continued)*

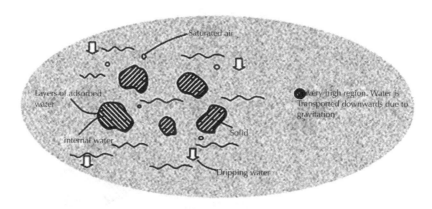

FIGURE 2 (a) Cross sections of nonabsorbent material at different regains. (b) Cross sections of absorbent material at different regains.

Fiber sorption properties influence the heat and mass transfer up to the point when the rate of increased conductivity with regain is low in the curves, and then all fiber types behave similarly. Generally heat transfer increases with increasing regain, but in this initial regain the rise is most pronounces for the nonabsorbent polypropylene. The fiber with the lowest effective conductivity over the regain 0–200% regain is wool, an effect that is especially pronounced in the region of low regains from zero to saturation. This is mostly influenced by fiber sorption properties. Low regains are most common in real wear situation. This is mostly influenced by fiber sorption properties. Low regains are most common in real wear situations.

This explains the popular association between wool and warmth in situations such as yachting, where the garment will very likely become wet. Cotton fabric has the highest effective thermal conductivity for almost the whole regain.

11.5.3 TRANSPORT PHENOMENA FOR SWEAT

Fabrics to protect human body are, in most cases used under non-equilibrium conditions; therefore, characteristics of fabrics under non-isothermal and non-equilibrium conditions are important in evaluating overall performance. Furthermore, in colder environments, layered fabrics rather than a single fabric are used in most cases. Under such conditions, the two most important characteristics of fabrics are water vapor and heat transport. However, the water vapor transport may not influenced significantly by surface characteristics-the hydrophilic or hydrophobic nature of fabrics. On the other hand, when liquid water contacted a fabric, such as in the case of sweating, the surface wet-ability of fabric play a dominant role in determining the water vapor transport through layered fabrics.

In such a case, the wicking characteristics, which determines how quickly and how widely liquid water spreads out laterally on the surface of or within the matrix of the fabric, determines the overall water vapor transport rate through the layered fabrics.

It should be noted that the overall water vapor and heat transport characteristics of a fabric should depend on other factors such as the water vapor absorbability of the fibers, the porosity, density, and thickness of the fabric, *etc*.

Moreover, transport phenomena for the sweat case are much more complicated than the water vapor case because wetting of the surface by liquid water precedes water wetting of the surface by liquid water precedes water vapor transmission. Note that there is an important difference in water absorbing characterizing of wool and cotton, although both fibers have relatively high water vapor absorption rates. Because of the hydrophobic surface of wool fibers, a liquid droplet in contact with a wool fabric does not spread out laterally within a fabric layer. The water vapor transport rate, in the sweat case, can be indicated by the size of liquid water spread out on the surface or within a fabric matrix.

Moreover, the term "breathable" implies that the fabric is actively ventilated. This is not the case. Breathable fabrics passively allow water vapor to diffuse through them yet still prevent the penetration of liquid water. Production of water vapor by the skin is essential for maintenance of body temperature. The normal body core temperature is 37°C, and skin temperature is between 33 and 35°C, depending on conditions. If the core temperature goes beyond critical limits of about 24°C and 45°C then death results. The narrower limits of 34°C and 42°C can cause adverse effects such as disorientation and convulsions. If the sufferer is engaged in a hazardous pastime or occupation then this could have disastrous consequences.

11.5.4 FACTORS INFLUENCING THE COMFORT ASSOCIATED WITH WEARING FABRICS

As it was mentioned earlier, if liquid water (sweat) cannot be dissipated quickly, the humidity of the air in the space in between the skin and the fabric that contacts with the skin rises. This increased humidity prevents rapid evaporation of liquid water on the skin and gives the body the sensation of "heat" that triggered the sweating in the first place. Consequently, the body responds with increased sweating to dissipate excess thermal energy. Thus a fabric's inability to remove liquid water seems to be the major factor causing uncomfortable feeling for the wearer.

Scientists conducted wearer trails for shirts made of various fibers.

They concluded that the largest factor that influence-wearing comfort was the ability of fibers to absorb water, regardless of whether fibers were synthetic or natural. All of these studies indicate that the transient state phenomenon responding to the physiological demand to cause sweating is most relevant to comfort or discomfort associated with fabrics. It is important to point out that a highly water absorbing fabric placed in the first layer keeps the partial pressure of water vapor near the skin low, which helps to dissipate water at the skin surface, although the vapor transport rate is smaller than for non-absorbing fabrics. In other words, the dissipation of water by means of absorption by fabrics appears to be much more efficient way to keep water vapor pressure near the skin low than dissipation by permeation through fabrics. Highly water-absorbing fabrics raise the temperature of the air space near the skin. The temperature rise will further decrease relative humidity; however, the higher temperature may or may not be desirable depending on environmental conditions.

11.5.5 INTERACTION OF MOISTURE WITH FABRICS

To stay warm and dry while active outdoors in winter has always been a challenge. In the worst case, an individual exercises strenuously, sweats profusely, and then rests. During exercise, liquid water accumulates on the skin and starts to wet the clothing layers above skin. Some of the sweat evaporates from both the skin and the clothing. Depending on the temperature and humidity gradient across the clothing, the water vapor either leaves the clothing or condenses and freezes somewhere in its outer layers.

When one stops exercising and begins to rest, active sweating soon ceases. This allows the skin and clothing layers eventually dry. During this time, however, the heat loss from body can be considerable. Heat is taken from the body to evaporate the sweat, both that on the skin and that in the clothing. The heat flow from the skin through the clothing can be considerably greater when the clothing is very wet, since water decreases clothing's thermal insulation. This post-exercise chill can be exceedingly comfortable and can lead to dangerous hypothermia.

A dry layer next to the skin is more comfortable than a wet one. If one can wear clothing next to the skin that does not pick up any moisture, but rather passes it through to a layer away from the skin, heat loss at rest will be reduced. For such reasons, synthetic fibers have gained popularity with winter enthusiasts such as hikers and skiers.

Advertising the popular press would have us believe that synthetic materials pick up very little moisture, dry quickly, and so leave the wearer warm and dry. In contrast, warnings are given against wearing cotton or wool next to skin, since these fibers absorb sweat and so "lower body temperature." A further property credited to synthetics, in particular polypropylene, is that they wick water away from the skin, leaving one dry and comfortable.

In the early fifties, when synthetic fibers such as nylon and the acrylics were first coming onto the consumer market, Scientists compared the water absorption and drying properties of these "miracle" fibers with those of conventional wool and cotton. Forty-five years latter, the water absorption and drying properties of synthetics were compared with natural fibers and it was found that all fabrics pick up water, and the time they take to dry is proportional to the amount of water they initially pick up.

It was also found that properties relevant to clothing on an exercising person, that is, the energy required to evaporated water from under and through a dry fabric or to dry a wet fabric and layer-to-layer wicking.

Researches compared the heat exchange and thermal insulation of two ensembles, one made from wool, the other from nylon, worn by subjects who exercised either lightly (dry condition) or strenuously (wet condition) for 60 minutes, then rested 60 minutes.

He found that there was a significant difference in the physiological and subjective responses between dry and wet conditions, but not between the two fiber types. Further, there was no significant difference between the ratings of temperature and humidity sensations for the wool and nylon garments. The wool garment picked up more water than the nylon garment (245 g versus 198 g) for the wet condition. However, the wool fabric may have been slightly thicker than the nylon fabric, since it was reported to have a slightly greater thermal resistance and would therefore, holding more water.

11.5.6 MOISTURE TRANSFER IN TEXTILES

In nude man any increase of sweating is immediately accompanied by an increase in heat loss due to evaporation. Similarly any decrease in sweating is immediately accompanied by a decrease in heat loss. Thus, nude man has a control of his heat loss, which has no appreciable time lag. This is shown diagrammatically in Fig. 3 of Section 11.5.

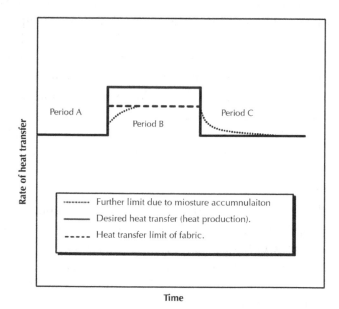

FIGURE 3   Rate of heat transfer versus time.

In Fig. 3 of Section 11.5, time is plotted as abscissa and rate of heat produced or lost as ordinate. To maintain perfect heat balance and a constant temperature, heat loss should equal heat production so that the heat production and heat loss curves should be the same. Suppose a man is initially at rest with a low heat production and a like heat loss as represented by the solid line in period A. When he exercises and produce more heat, the heat loss should rise as represented by the solid line in period B. Again, when he returns to the resting condition, period C, heat loss should return to the solid base line.

If sweating is the mechanism bringing about increasing heat loss but evaporation is limited, the increased heat loss might only be sufficient to match an increased heat production represented by the dashed line in period B. The position of this dashed line will depend on the equilibrium vapor transfer characteristics of the clothing. If, however, the hypothetical man is clothed in absorbent clothing, some of the sweat initially evaporated at the skin at the start of the exercise period will be absorbed by the clothing and its heat of absorption will appear in the clothing as sensible heat. This source of sensible heat will temporarily reduce the heat loss so that it follows the dotted line. Eventually a new equilibrium moisture content will be established and the dotted and dashed lines will coincide. When exercise and sweating stop, period C, moisture accumulated in the clothing will be desorbed or evaporated and tend to cool the clothing and the man wearing it. Thus, there is a time lag, and the heat loss curve will tend to follow the dotted curve during the after-exercise period. Since in Fig.3 heat loss per unit time is plotted against time, the area between the dotted line and the solid line represents an amount of heat, as distinguished from rate of heat loss, which can be regarded as a quantitative value of after exercise chill.

It should be noted that the moisture contained in the clothing need not be only that which is collected by absorption. It is also possible in cold damp or extreme cold environments that sweat, which is evaporated at the skin, will re-condense when it reaches colder layers of clothing. Alternatively the sweat rate may be so high that some of it will not evaporate from the skin. In nude man this drips off, but in clothed man it is blotted up by clothing to evaporate after sweating ceases.

Meanwhile, measurements of water vapor permeability of woven fabrics have indicated that in the lower ranges of fabric density, the main path of water vapor transfer is through the air spaces between fibers and yarns. This covers the densities characteristic of most apparel fabrics made from staple fibers, although filament yarn fabrics may be woven to higher densities in which the kind of fiber itself in the passage of water vapor, it is necessary to account for the water vapor passage through air spaces.

11.5.7 WATER VAPOR SORPTION MECHANISM IN FABRICS

Scientists proposed a mathematical model for describing heat and moisture transfer in fabric, as shown in Eqs. (1) and (2) of Section 11.5, and the further analyzed the model in 1948 [64].

$$\varepsilon \frac{\partial C_a}{\partial t} + (1-\varepsilon)\frac{\partial C_f}{\partial t} = \frac{D_a \varepsilon}{\tau}\frac{\partial^2 C_a}{\partial x^2} \tag{1}$$

$$C_v \frac{\partial T}{\partial t} - \lambda \frac{\partial C_f}{\partial t} = K \frac{\partial^2 T}{\partial x^2} \tag{2}$$

In these equations, both C_v and λ are functions of the concentration of water absorbed by the fibers. Most textile fibers have very small diameters and very large surface/volume ratios. The assumption in the second equation of instantaneous thermal equilibrium between the fibers and the gas in the inter-fiber space does not therefore lead to appreciable error. The two equations in the model are not linear and contain three unknown, i.e., C_f, T, and C_a. A third equation should be established appropriately in order to solve the model. Researchers derived a third equation to obtain an analytical solution by assuming that C_f is linearly dependent on T and C_a, and that fibers reach equilibrium with adjacent air instantaneously. Considering the two-stage sorption process of wool, Scientists proposed an exponential relationship to describe the rate of water content change in the fibers, as shown in:

$$\frac{1}{\varepsilon}\frac{\partial C_f}{\partial t}(H_a - H_f)\gamma \tag{3}$$

where

$$\gamma = k_1(1 - \exp(k_2|H_a - H_f|) \tag{4}$$

and k_1 and k_2 are adjustable parameters that are evaluated by comparing the prediction of the model and measured moisture content of the fabric.

Scientists reported a numerical model describing combined heat and water vapor transport through clothing.

The assumptions in his model do not allow for the complexity of the moisture sorption isotherm and the sorption kinetics of fibers. Scientists presented two mathematical models to simulate the interaction between moisture sorption by fiber and moisture flux through the air spaces of a fabric. In the first model, they considered diffusion within the fiber to be so rapid that the fiber moisture content is always in equilibrium with the adjacent air. In the second model, they assumed that the sorption kinetics of the fiber follows Fickian diffusion. Their model neglected the effect of heat of sorption behavior of the fiber. Scientists developed a new sorption equation that takes into account the two-stage sorption kinetics of wool fibers, and incorporated this with more realistic boundary conditions to simulate the sorption behavior of wool fabrics. They assumed that water vapor uptake rate of fiber consists of a two components associated with the two stages of sorption identified by researchers.

The first stage is represented by Fickian diffusion with a constant coefficient. Second-stage sorption is much slower than the first and follows an exponential relationship. The relative contributions of the two stages to the total uptake vary with the sorption stage and the initial regain of the fibers. Thus, the sorption rate equation can be written as:

$$\frac{\partial C_f}{\partial t}(1-p)R_1 + pR_2 \tag{5}$$

where R_1 is the first-stage sorption rate, R_2 is the second-stage rate sorption rate, and p is a proportional of uptake in the second stage. Eq. (5) of Section 11.5 assumes that the sorption rate is a linear average of the first and second sorption rates. The first-stage sorption rate R_1 can be derived using Crank's truncated solution.

This may lead to a corresponding algorithm that needs a strict time restriction and hence long computation times.

The second-stage sorption rate R_2, which relates local temperature, humidity, and the sorption history of the fabric, is assumed to have the following form:

$$R_2(x,t) = s_1 sign(H_a(x,t) - H_a(x,t) - H_f(x,t)$$

$$\times \exp\left(\frac{s_2}{|H_a(x,t) - H_f(x,t)|}\right) \tag{6}$$

where s_1 and s_2 are constants. No values for s_1 and s_2 have been reported in the literature for any textile fibers. This is also an empirical equation that has an unclear physical meaning, which makes it inconvenient to predict and simulate heat and moisture transport in a fabric. These equations were improved substantially by the scientists. The numerical values and approximate relationships they used are listed in Table 3 of Section 11.5. They assumed that moisture sorption by a wool fiber can be generally described by a uniform diffusion equation for both stages of sorption.

$$\frac{\partial C_f(x,r,t)}{\partial t} = \frac{1}{r}\frac{\partial}{\partial r}\left(rD_f(x,t)\frac{\partial C_f(x,r,t)}{\partial r}\right)$$

$$C_{fs}(x,R_f,t) = f(H_a(x,t),T(x,t)) \tag{7}$$

TABLE 3 Numerical values of wool and physical properties.

Parameters	Initial values	Mathematical relationship
Thermal conductivity of fabric $(KJ/m.K)$	$3.8493\mathrm{e}^{-2}$	$(38.493 - 0.72W_c + 0.113W_c^2$ $-0.002W_c^3)10^{-3}$
Volumetric heat capacity of fabric $(kJ/m^3.K)$	1609.7	$373.3 + 4661\ W_C + 4.221\ T$
Diffusion coefficient of fiber (m^2/s)	$2.4435\ e^{-14}$	$1.0637\ \mathrm{arc\ tan}(1541.1933)$ $(3600/t^2)10^{-14}$
Diffusion coefficient of water vapor in fabric (m^2/s)	$1.91\ e^{-5}$	_____
Heat of sorption or adsorption of water by fibers (KJ/Kg)	4124.5	$1602.5\exp(-11.72\ W_c) + 2522$
Porosity of fabric	0.925	_____
Density of fabric (Kg/m^3)	1330	_____
Radius of wool fiber (m)	$1.04e^{-5}$	_____
Mass transfer coefficient (m/s)	0.137	_____
Heat transfer coefficient $(W/m^2.K)$	99.4	_____

W_c = Water content of the fibers in the fabric.

where $D_f(x,t)$ are the diffusion coefficients that have different presentations at different stages of sorption, and x is the coordinate of a fiber in the given fabric. The boundary condition is determined by the relative humidity of the air surrounding a fiber at x. In wool fabric, $D_f(x,t)$ is a function of $W_c(x,t)$, which depends on the sorption time and the fiber location.

11.5.8 MODELING

The fabric model simulates the transport of a liquid and vapor-phase fluid that can undergo phase change (e.g., water) and an inert gas (air) in a textile layer. Several new models and capabilities were added to a standard commercial CFD code (FLUENT Version 6.0, Fluent Inc., Lebanon, NH) [47]. These capabilities include:

- Vapor phase transport (variable permeability).
- Liquid phase transport (wicking).
- Fabric property dependence on moisture content.
- Vapor/liquid phase change (evaporation/condensation).
- Sorption to fabric fibers.

In the fabric, transport equations are derived for mass, momentum, and energy in the gas and liquid phases by volume-averaging techniques. Definitions for intrinsic phase average, global phase average, and spatial average for porous media are those given by scientists. Since the fabric porosity is not constant due to changing amounts of liquid and bound water, the source term for each transport equation includes quantities that arise due to the variable porosity. These equations are summarized in general form below .

Gas phase continuity equation:

$$\frac{\partial}{\partial t}\left(\left(1-\varepsilon_{ds}\right)\rho_{\gamma}\right)+\nabla.\left(\rho_{\gamma}v_{\gamma}\right)=S_{\gamma} \tag{8}$$

$$S_{y}=m'_{sv}+m'_{lv}+\frac{\partial}{\partial t}\left(\left(\varepsilon_{bl}+\varepsilon_{\beta}\right)\rho_{\gamma}\right) \tag{9}$$

Vapor continuity equation:

$$\frac{\partial}{\partial t}\left(\left(1-\varepsilon_{ds}\right)\rho_{\gamma}m_{v}\right)+\nabla.\left(\rho_{\gamma}m_{v}v_{\gamma}\right)=\nabla.\left\{\rho_{\gamma}D_{eff}\nabla\left(m_{v}\right)\right\}+S_{v} \tag{10}$$

$$S_v = m''^m_{sv} + m''^m_{lv} + \frac{\partial}{\partial t}\{(\varepsilon_b + \varepsilon_\beta)\rho_\gamma m_v\} \qquad (11)$$

$$m_v = \frac{\rho_v}{\rho_\gamma} \qquad (12)$$

Gas phase momentum equation:

$$\frac{\partial}{\partial t}(\rho_\gamma v_\gamma) + \nabla.(\rho_\gamma v_\gamma) = \nabla.\{\mu_\gamma \nabla(v_\gamma)\} - \nabla p_\gamma + S_\gamma \qquad (13)$$

$$S_\gamma = -v_\gamma \frac{\mu_\gamma}{K_\gamma k_\gamma} \qquad (14)$$

Liquid transport:

$$\frac{\partial}{\partial t}\left[(1-\varepsilon_{ds})\rho_\gamma s\right] = \nabla.\left(-\frac{k_\beta K_\beta}{\mu_\beta}\frac{\partial P_c}{\partial s}\right)\nabla s + S_l \qquad (15)$$

$$S_l = -\nabla.\left(\frac{k_\beta K_\beta}{\mu_\beta}\rho_\beta g\right) - \frac{m''^m_{ls}}{\rho_\beta} - \frac{m''^m_{lv}}{\rho_\beta} + \frac{\partial}{\partial t}\left[(\varepsilon_{bl})s\right] +$$

$$\frac{\partial}{\partial t}\left[(1-\varepsilon_{ds})(\rho_\gamma - 1)s\right] \qquad (16)$$

$$\frac{\partial}{\partial t}\left[(1-\varepsilon_{ds})\rho_\gamma h_\gamma + \varepsilon_{ds}\rho_{ds} h_{ds}\right] + \nabla.(\rho_\gamma v_\gamma h_\gamma) + \nabla.(\rho_\beta v_\beta (h_v - \Delta h_v))$$

$$= \nabla . \left[k_{\mathit{eff}} \nabla (T) \right] + \nabla . \left(-h_a J_a - h_v J_v \right) + S_T \tag{17}$$

$$S_T = \frac{\partial}{\partial t} \left[\left(\varepsilon_\beta + \varepsilon_{bl} \right) \rho_\gamma h_\gamma - \varepsilon_\beta \rho_\beta \left(h_v - \Delta h_v \right) - \varepsilon_{bl} \rho_\beta \left(h_v - \Delta h_v - Q \right) \right] \tag{18}$$

$$v_\beta = \frac{k_\beta K_\beta}{\mu_\beta} \left[\frac{\partial P_c}{\partial s} \nabla s + \rho_\beta g \right] \tag{19}$$

In summery, due to the intensive body activity, the wearer perspires and the cloth worn next to skin will get wet. These moisture fabrics reduce the body heat and make the wearer to become tired. So the cloth worn next to the skin should assist for the moisture release quickly to the atmosphere. The fabric worn next to the skin should have two important properties. The initial and fore most property is to evaporate the perspiration from the skin surface and the second property is to transfer the moisture to the atmosphere and make the wearer to feel comfort. Diffusion and wicking are the two ways by which the moisture is transferred to the atmosphere. These two are mostly governed by the fiber type and fabric stricture. The airflow through the fabric makes the moisture to evaporate to the atmosphere. The capillary path plays a vital role in the transfer of moisture and this depends on the wicking behavior of the fabric. In the development of protective clothing and other textiles, modeling offers a powerful companion to experiments and testing.

It should be noted that condensation occurs when the vapor density of the steam is higher that its saturation vapor density. The condensation rate is proportional to the vapor density difference between that in the gas phase and that at the condensing surface. The relative hygrometry is quite different due to the action (water vapor pressure) of the airs on each side of the fabric (Fig. 4 of Section 11.5). Sorption and adsorption have not opposite kinetics, the former faster in temperature and in charge of humidity during the first minutes, the latter more complete in discharge of humidity after a long time (Fig. 5 of Section 11.5). Figure 6 of Section 11.5 shows experimental results of hygrometry for the internal and external surfaces of wool fabric (between skin and fabric). Meanwhile, the internal gap has a considerable effect on the moisture transmission rate. The internal air gap has been identified as being a source of potential errors in most experimental works due to its changing resistance (Fig. 7 of Section 11.5). Figure 8 of Section 11.5 shows the effect of thickness on the amount of water can be held in a fabric.

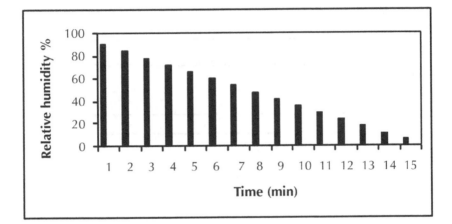

FIGURE 4   Relative hygrometry of wool fiber during adsorption.

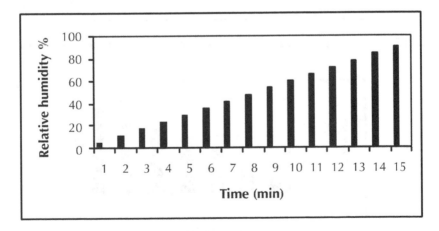

FIGURE 5   Relative hygrometry of wool fiber during sorption.

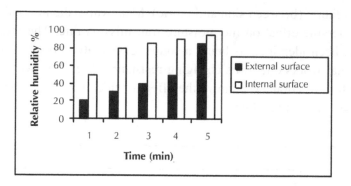

FIGURE 6   Experimental results of hygrometry for the internal and external surfaces of wool fabric (between skin and fabric).

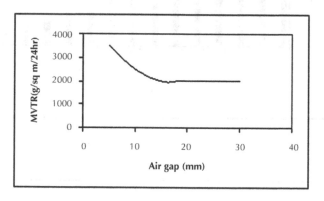

FIGURE 7   Effect of the internal air gap size on the moisture vapor transmission (MVTR).

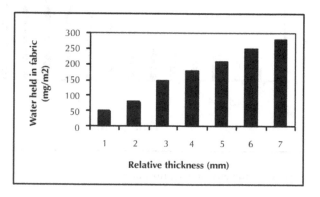

FIGURE 8   Thickness versus the amount of water held in fabric.

11.6 APPENDIX

11.6.1 HEAT TRANSFER

Heat transfer is the study of energy movement in the form of heat, which occurs in many types of processes. The transfer occurs from the high to the low temperature regions. Therefore a temperature gradient has to exist between the two regions for heat transfer to happen. It can be done by conduction (within one solid or between two solids in contact), by convection (between two fluids or a fluid and a solid in direct contact with the fluid), by radiation (transmission by electromagnetic waves through space) or by combination of the above three methods.

The general equation for heat transfer is:

$$\begin{pmatrix} \text{rate of} \\ \text{heat in} \end{pmatrix} + \begin{pmatrix} \text{rate of generation} \\ \text{of heat} \end{pmatrix} = \begin{pmatrix} \text{rate of} \\ \text{heat out} \end{pmatrix} + \begin{pmatrix} \text{rate of accumulation} \\ \text{of heat} \end{pmatrix}$$

11.6.2 MASS TRANSFER

Textiles exposed to a hot air stream may be cooled evaporative by bleeding water through its surface. Water vapor may condense out of damp air onto cool surfaces. Heat will flow through an air-water mixture in these situations, but water vapor will diffuse through air as well. This sort of transport of one substance relative to another called mass transfer.

11.6.3 DRYING OF TEXTILES

Drying of textiles is accomplished by vaporizing the water and to do this the latent heat of vaporization must be supplied. There are, thus, two important process-controlling factors that enter into the process of drying:

- Transfer of heat to provide the necessary latent heat of vaporization,

- Movement of water or water vapor through textiles and then away from it to effect separation of water.

Drying processes fall into different categories:

In air and contact drying, heat is transferred through the textiles either from heated air or from heated surfaces.

The water vapor is removed with the air. Heat transfer in is generally by convection, conduction, sometimes by radiation.

11.6.4 STEADY STATE HEAT TRANSFER

Heat transfer is said to be at steady-state when the quantity of heat flowing from one point to another by unit time is constant and the temperatures at each point in the system do not change with time.

Assuming no heat generation, no accumulation of heat and transfer of heat by conduction, at steady state we have:

$$q_x = q_{x+\Delta x}$$

$$q_x = -kA\frac{dT}{dx}$$

q_x / A is the heat flux in Wm^{-2} while the quantity dT/dx represent the temperature gradient in the x direction.

11.6.5 CONVECTIVE HEAT-TRANSFER COEFFICIENT

When a fluid is in forced or natural convective motion along a surface, the rate of heat transfer between the solid and the fluid is expressed by the following equation:

$$q = h.A(T_W - T_f)$$

The coefficient h is dependent on the system geometry, the fluid properties and velocity and the temperature gradient. Most of the resistance to heat transfer happens in the stationary layer of fluid present at the surface of the solid, therefore the coefficient h is often called film coefficient.

11.6.6 DIMENSIONLESS NUMBERS IN CONVECTIVE HEAT TRANSFER

Correlations for predicting film coefficient h are semi empirical and use dimensionless numbers, which describe the physical properties of the fluid, the type of flow, the temperature difference and the geometry of the system.

The Reynolds Number characterizes the flow properties (laminar or turbulent). L is the characteristic length: length for a plate, diameter for cylinder or sphere.

$$N_{Re} = \frac{\rho L v}{\mu}$$

The Prandtl Number characterizes the physical properties of the fluid for the viscous layer near the wall.

$$N_{Pr} = \frac{\mu c_p}{k}$$

The Nusselt Number relates the heat transfer coefficient h to the thermal conductivity k of the fluid.

$$N_{Nu} = \frac{hL}{k}$$

The Grashof Number characterizes the physical properties of the fluid for natural convection.

$$N_{Gr} = \frac{L^3 \Delta \rho g}{\rho \gamma^2} = \frac{L^3 \rho^2 g \beta \Delta T}{\mu^2}$$

11.6.7   RADIATION HEAT TRANSFER

Radiation is a term applied to many processes which involve energy transfer by electromagnetic wave (x rays, light, gamma rays, etc.). It obeys the same laws as light, travels in straight lines and can be transmitted through space and vacuum. It is an important mode of heat transfer encountered where large temperature difference occurs between two surfaces such as in furnaces, radiant driers and baking ovens.

The thermal energy of the hot source is converted into the energy of electromagnetic waves. These waves travel through space into straight lines and strike a cold surface. The waves that strike the cold body are absorbed by that body and converted back to thermal energy or heat. When thermal radiations falls upon a body, part is absorbed by the body in the form of heat, part is reflected back into space and in some case part can be transmitted through the body.

The basic equation for heat transfer by radiation from a body at temperature T is:

$$q = A \varepsilon \sigma T^4$$

where ε is the emissivity of the body. $\varepsilon = 1$ for a perfect black body while real bodies which are gray bodies have an $\varepsilon < 1$.

11.6.8   COMBINING RADIATION AND CONVECTION HEAT TRANSFER

In many applications radiation and convection occurs at the same time. A body receiving energy from radiation will return some to its surrounding (unless in vacuum) through convective heat transfer.

For convenience, a radiation heat transfer coefficient h_R expressed in $W/m^2.K$ can be evaluated in the following manner:

$$q = A\varepsilon\sigma\left(T_1^4 - T_2^4\right) = h_r A\left(T_1 - T_2\right)$$

$$h_r = \varepsilon\left(5.67\times10^{-8}\right)\frac{\left(T_1^4 - t_2^4\right)}{\left(T_1 - T_2\right)}$$

11.6.9 CONDUCTIVE HEAT TRANSFER

Fourier's Law can be integrated through a flat wall of constant cross section A for the case of steady-state heat transfer when the thermal conductivity of the wall k is constant.

$$\frac{q}{A}\int_{x_1}^{x_2} dx = -k\int_{T_1}^{T_2} dT \rightarrow \frac{q}{A} = \frac{k}{\Delta x}\left(T_1 - T_2\right)$$

At any position x between x_1 and x_2, the temperature T varies linearly with the distance:

$$\frac{q}{A} = \frac{k}{x - x_1}\left(T_1 - T\right)$$

11.6.10 GENERAL CONCEPTS

For heat flow analysis of wet porous nanostructure fabrics, the liquid is water and the gas is air. Evaporation or condensation occurs at the interface between the water and air so that the air is mixed with water vapor. A flow of the mixture of air and vapor may be caused by external forces, for instance, by an imposed pressure difference. The vapor will also move relative to the gas by diffusion from regions where the partial pressure of the vapor is higher to those where it is lower.

Heat flow in porous nanostructure fabrics is the study of energy movement in the form of heat, which occurs in many types of processes. The transfer of heat in porous nanostructure fabrics occurs from the high to the low temperature regions. Therefore a temperature gradient has to exist between the two regions for heat transfer to happen. It can be done by conduction (within one porous solid or between two porous solids in contact), by convection (between two fluids or a fluid and a porous solid in direct contact with the fluid), by radiation (transmission by electromagnetic waves through space) or by combination of the above three methods.

The general equation for heat transfer in porous media is:

$$\begin{pmatrix} \text{rate of} \\ \text{heat in} \end{pmatrix} + \begin{pmatrix} \text{rate of generation} \\ \text{of heat} \end{pmatrix} = \begin{pmatrix} \text{rate of} \\ \text{heat out} \end{pmatrix} + \begin{pmatrix} \text{rate of accumulation} \\ \text{of heat} \end{pmatrix}$$

When a wet porous nanostructure fabrics material is subjected to thermal drying two processes occur simultaneously, namely:
- Transfer of heat to raise the wet porous media temperature and to evaporate the moisture content.
- Transfer of mass in the form of internal moisture to the surface of the porous material and its subsequent evaporation.

The rate at which drying is accomplished is governed by the rate at which these two processes proceed. Heat is a form of energy that can across the boundary of a system. Heat can, therefore, be defined as "the form of energy that is transferred between a system and its surroundings as a result of a temperature difference." There can only be a transfer of energy across the boundary in the form of heat if there is a temperature difference between the system and its surroundings. Conversely, if the system and surroundings are at the same temperature there is no heat transfer across the boundary.

Strictly speaking, the term "*heat*" is a name given to the particular form of energy crossing the boundary. However, heat is more usually referred to in thermodynamics through the term "heat transfer," which is consistent with the ability of heat to raise or lower the energy within a system.

There are three modes of heat flow in porous nanostructure fabrics media:
- Convection

- Conduction
- Radiation

All three are different. Convection relies on movement of a fluid in porous material. Conduction relies on transfer of energy between molecules within a porous solid or fluid. Radiation is a form of electromagnetic energy transmission and is independent of any substance between the emitter and receiver of such energy. However, all three modes of heat flow rely on a temperature difference for the transfer of energy to take place.

The greater the temperature difference the more rapidly will the heat be transferred. Conversely, the lower the temperature difference, the slower will be the rate at which heat is transferred. When discussing the modes of heat transfer it is the rate of heat transfer Q that defines the characteristics rather than the quantity of heat.

As it was mentioned earlier, there are three modes of heat flow in porous structures, convection, conduction and radiation. Although two, or even all three, modes of heat flow may be combined in any particular thermodynamic situation, the three are quite different and will be introduced separately.

The coupled heat and liquid moisture transport of porous material has wide industrial applications. Heat transfer mechanisms in porous textiles include conduction by the solid material of fibers, conduction by intervening air, radiation, and convection. Meanwhile, liquid and moisture transfer mechanisms include vapor diffusion in the void space and moisture sorption by the fiber, evaporation, and capillary effects. Water vapor moves through porous textiles as a result of water vapor concentration differences. Fibers absorb water vapor due to their internal chemical compositions and structures. The flow of liquid moisture through the textiles is caused by fiber-liquid molecular attraction at the surface of fiber materials, which is determined mainly by surface tension and effective capillary pore distribution and pathways. Evaporation and/or condensation take place, depending on the temperature and moisture distributions. The heat transfer process is coupled with the moisture transfer processes with phase changes such as moisture sorption/desorption and evaporation/condensation.

11.6.11 DRYING OF POROUS NANOSTRUCTURE FABRICS

All three of the mechanisms by which heat is transferred- conduction, radiation and convection, may enter into drying. The relative importance of the mechanisms varies from one drying process to another and very often one mode of heat transfer predominates to such extent that it governs the overall process.

As an example, in air-drying the rate of heat transfer is given by Eq. (1) of Appendix:

$$q = h_s A(T_a - T_s)$$
(1)

Where q is the heat transfer rate in Js^{-1}, h_s is the surface heat-transfer coefficient in Jm^{-2} s^{-1} °C^{-1}, A is the area through which heat flow is taking place, m^{-2}, T_a is the air temperature and T_s is the temperature of the surface which is drying, °C.

To take another example, in a cylindrical dryer where moist material is spread over the surface of a heated cylinder, heat transfer occurs by conduction from the cylinder to the porous media, so that the Eq. (2) of Appendix is:

$$q = UA(T_i - T_s)$$
(2)

Where U is the overall heat-transfer coefficient, T_i is the cylinder temperature (usually very close to that of the steam), T_s is the surface temperature of textile and A is the area of the drying surface on the cylinder. The value of U can be estimated from the conductivity of the cylinder material and of the layer of porous solid.

Mass transfer in the drying of a wet porous material will depend on two mechanisms: movement of moisture within the porous material which will be a function of the internal physical nature of the solid and its moisture content; and the movement of water vapor from the material surface as a result of water vapor from the material surface as a result of external conditions of temperature, air humidity and flow, area of exposed surface and supernatant pressure.

Some porous materials such as textiles exposed to a hot air stream may be cooled evaporative by bleeding water through its surface. Water vapor may condense out of damp air onto cool surfaces. Heat will flow through an air-water mixture in these situations, but water vapor will diffuse through air as well. This sort of transport of one substance relative to another called mass transfer. The moisture content, X, is described as the ratio of the amount of water in the materials, m_{H_2O} to the dry weight of material, $m_{material}$:

$$X = \frac{m_{H_2O}}{m_{material}} \qquad (3)$$

There are large differences in quality between different porous materials depending on structure and type of material. A porous material such as textiles can be hydrophilic or hydrophobic. The hydrophilic fibers can absorb water, while hydrophobic fibers do not. A textile that transports water through its porous structures without absorbing moisture is preferable to use as a first layer. Mass transfer during drying depends on the transport within the fiber and from the textile surface, as well as on how the textile absorbs water, all of which will affect the drying process.

As the critical moisture content or the falling drying rate period is reached, the drying rate is less affected by external factors such as air velocity. Instead, the internal factors due to moisture transport in the material will have a larger impact. Moisture is transported in textile during drying through:

- Capillary flow of unbound water;
- Movement of bound water; and
- Vapor transfer.

Unbound water in a porous media such as textile will be transported primarily by capillary flow.

As water is transported out of the porous material, air will be replacing the water in the pores. This will leave isolated areas of moisture where the capillary flow continues.

Moisture in a porous structure can be transferred in liquid and gaseous phases. Several modes of moisture transport can be distinguished:

- Transport by liquid diffusion;

- Transport by vapor diffusion;
- Transport by effusion (Knudsen-type diffusion);
- Transport by thermo-diffusion;
- Transport by capillary forces;
- Transport by osmotic pressure; and
- Transport due to pressure gradient.

11.6.12   CONVECTION HEAT FLOW IN POROUS MEDIA

A very common method of removing water from porous structures is convective drying. Convection is a mode of heat transfer that takes place as a result of motion within a fluid. If the fluid, starts at a constant temperature and the surface is suddenly increased in temperature to above that of the fluid, there will be convective heat transfer from the surface to the fluid as a result of the temperature difference. Under these conditions the temperature difference causing the heat transfer can be defined as:

ΔT = surface temperature-mean fluid temperature

Using this definition of the temperature difference, the rate of heat transfer due to convection can be evaluated using Newton's law of cooling:

$$Q = h_c A \Delta T \tag{4}$$

where A is the heat transfer surface area and h_c is the coefficient of heat transfer from the surface to the fluid, referred to as the "convective heat transfer coefficient."

The units of the convective heat transfer coefficient can be determined from the units of other variables:

$$Q = h_c A \Delta T$$
$$W = (h_c) m^2 K \tag{5}$$

so the units of h_c are $W / m^2 K$.

The relationships given in Eqs. (4) and (5) of Appendix are also true for the situation where a surface is being heated due to the fluid having higher temperature than the surface. However, in this case the direction of heat transfer is from the fluid to the surface and the temperature difference will now be

ΔT = mean fluid temperature-surface temperature

The relative temperatures of the surface and fluid determine the direction of heat transfer and the rate at which heat transfer take place.

As given in previous equations, the rate of heat transfer is not only determined by the temperature difference but also by the convective heat transfer coefficient h_c. This is not a constant but varies quite widely depending on the properties of the fluid and the behavior of the flow. The value of h_c must depend on the thermal capacity of the fluid particle considered, i.e., mC_p for the particle.

Two common heat transfer fluids are air and water, due to their widespread availability. Water is approximately 800 times denser than air and also has a higher value of C_p. If the argument given above is valid then water has a higher thermal capacity than air and should have a better convective heat transfer performance. This is borne out in practice because typical values of convective heat transfer coefficients are as follows:

Fluid	$h_c \left(W / m^2 K \right)$
Water	500–10,000
Air	5–100

The variation in the values reflects the variation in the behavior of the flow, particularly the flow velocity, with the higher values of h_c resulting from higher flow velocities over the surface.

When a fluid is in forced or natural convective motion along a surface, the rate of heat transfer between the solid and the fluid is expressed by the following equation:

$$q = h.A\left(T_W - T_f\right) \qquad (6)$$

The coefficient h is dependent on the system geometry, the fluid properties and velocity and the temperature gradient. Most of the resistance to heat transfer happens in the stationary layer of fluid present at the surface of the solid, therefore the coefficient h is often called film coefficient.

Correlations for predicting film coefficient h are semi empirical and use dimensionless numbers, which describe the physical properties of the fluid, the type of flow, the temperature difference and the geometry of the system.

The Reynolds Number characterizes the flow properties (laminar or turbulent). L is the characteristic length: length for a plate, diameter for cylinder or sphere.

$$N_{Re} = \frac{\rho L v}{\mu} \tag{7}$$

The Prandtl Number characterizes the physical properties of the fluid for the viscous layer near the wall.

$$N_{Pr} = \frac{\mu c_p}{k} \tag{8}$$

The Nusselt Number relates the heat transfer coefficient h to the thermal conductivity k of the fluid.

$$N_{Nu} = \frac{hL}{k} \tag{9}$$

The Grashof Number characterizes the physical properties of the fluid for natural convection.

$$N_{Gr} = \frac{L^3 \Delta \rho g}{\rho \gamma^2} = \frac{L^3 \rho^2 g \beta \Delta T}{\mu^2} \tag{10}$$

11.6.13 CONDUCTION HEAT FLOW IN POROUS MATERIALS

If a fluid could be kept stationary there would be no convection-taking place. However, it would still be possible to transfer heat by means of conduction. Conduction depends on the transfer of energy from one molecule to another within the heat transfer medium and, in this sense, thermal conduction is analogous to electrical conduction.

Conduction can occur within both porous solids and fluids. The rate of heat transfer depends on a physical property of the particular porous solid of fluid, termed its thermal conductivity k, and the temperature gradient across the porous medium. The thermal conductivity is defined as the measure of the rate of heat transfer across a unit width of porous material, for a unit cross-sectional area and for a unit difference in temperature.

From the definition of thermal conductivity k it can be shown that the rate of heat transfer is given by the equation:

$$Q = \frac{kA\Delta T}{x} \tag{11}$$

where ΔT is the temperature difference $T_1 - T_2$, defined by the temperature on the either side of the porous solid. The units of thermal conductivity can be determined from the units of the other variables:

$$Q = kA\Delta T \,/\, x$$
$$W = (k)m^2 K \,/\, m \tag{12}$$

so the unit of k are $W \,/\, m^2 K \,/\, m$, expressed as W/mK.

Fourier's Law can be integrated through a flat wall of constant cross section A for the case of steady-state heat transfer when the thermal conductivity of the wall k is constant.

$$\frac{q}{A}\int_{x_1}^{x_2} dx = -k\int_{T_1}^{T_2} dT \rightarrow \frac{q}{A} = \frac{k}{\Delta x}(T_1 - T_2) \tag{13}$$

At any position x between x_1 and x_2, the temperature T varies linearly with the distance:

$$\frac{q}{A} = \frac{k}{x - x_1}(T_1 - T)$$ (14)

11.6.14   RADIATION HEAT FLOW IN POROUS SOLIDS

The third mode of heat flow, radiation, does not depend on any medium for its transmission. In fact, it takes place most freely when there is a perfect vacuum between the emitter and the receiver of such energy. This is proved daily by the transfer of energy from the sun to the earth across the intervening space.

Radiation is a form of electromagnetic energy transmission and takes place between all matters providing that it is at a temperature above absolute zero. Infrared radiation form just part of the overall electromagnetic spectrum. Radiation is energy emitted by the electrons vibrating in the molecules at the surface of a porous body. The amount of energy that can be transferred depends on the absolute temperature of the porous body and the radiant properties of the surface.

A porous body that has a surface that will absorb all the radiant energy it receives is an ideal radiator, termed a "black body." Such a porous body will not only absorb radiation at a maximum level but will also emit radiation at a maximum level. However, in practice, porous bodies do not have the surface characteristics of a black body and will always absorb, or emit, radiant energy at a lower level than a black body.

It is possible to define how much of the radiant energy will be absorbed, or emitted, by a particular surface by the use of a correction factor, known as the "emissivity" and given the symbol ε. The emissivity of a surface is the measure of the actual amount of radiant energy that can be absorbed, compared to a black body. Similarly, the emissivity defines the radiant energy emitted from a surface compared to a black body. A black body would, therefore, by definition, have an emissivity ε of 1.

Since World War II, there have been major developments in the use of microwaves for heating applications. After this time it was realized that microwaves had the potential to provide rapid, energy-efficient heating of materials. These main applications of microwave heating today include food processing, wood drying, plastic and rubber treating as well as curing and preheating of ceramics. Broadly speaking, microwave radiation is the term associated with any electromagnetic radiation in the microwave frequency range of 300 MHz–300 Ghz. Domestic and industrial microwave ovens generally operate at a frequency of 2.45 Ghz corresponding to a wavelength of 12.2 cm. However, not all materials can be heated rapidly by microwaves. Porous materials may be classified into three groups, i.e., conductors insulators and absorbers. Porous materials that absorb microwave radiation are called dielectrics, thus, microwave heating is also referred to as dielectric heating. Dielectrics have two important properties:

- They have very few charge carriers. When an external electric field is applied there is very little change carried through the material matrix.
- The molecules or atoms comprising the dielectric exhibit a dipole movement distance.

An example of this is the stereochemistry of covalent bonds in a water molecule, giving the water molecule a dipole movement. Water is the typical case of non-symmetric molecule. Dipoles may be a natural feature of the dielectric or they may be induced. Distortion of the electron cloud around non-polar molecules or atoms through the presence of an external electric field can induce a temporary dipole movement. This movement generates friction inside the dielectric and the energy is dissipated subsequently as heat.

The interaction of dielectric materials with electromagnetic radiation in the microwave range results in energy absorbance. The ability of a material to absorb energy while in a microwave cavity is related to the loss tangent of the material.

This depends on the relaxation times of the molecules in the material, which, in turn, depends on the nature of the functional groups and the volume of the molecule. Generally, the dielectric properties of a material are related to temperature, moisture content, density and material geometry.

An important characteristic of microwave heating is the phenomenon of "hot spot" formation, whereby regions of very high temperature form due to non-uniform heating. This thermal instability arises because of the non-linear dependence of the electromagnetic and thermal properties of material on temperature. The formation of standing waves within the microwave cavity results in some regions being exposed to higher energy than others.

Cavity design is an important factor in the control, or the utilization of this "hot spots" phenomenon.

Microwave energy is extremely efficient in the selective heating of materials as no energy is wasted in "bulk heating" the sample. This is a clear advantage that microwave heating has over conventional methods. Microwave heating processes are currently undergoing investigation for application in a number of fields where the advantages of microwave energy may lead to significant savings in energy consumption, process time and environmental remediation.

Compared with conventional heating techniques, microwave heating has the following additional advantages:
- Higher heating rates;
- No direct contact between the heating source and the heated material;
- Delective heating may be achieved;
- Greater control of the heating or drying process;
- Reduced equipment size and waste.

As mentioned earlier, radiation is a term applied to many processes, which involve energy transfer, by electromagnetic wave (x-rays, light, gamma rays, etc.). It obeys the same laws as light, travels in straight lines and can be transmitted through space and vacuum. It is an important mode of heat transfer encountered where large temperature difference occurs between two surfaces such as in furnaces, radiant driers and baking ovens.

The thermal energy of the hot source is converted into the energy of electromagnetic waves. These waves travel through space into straight lines and strike a cold surface. The waves that strike the cold body are absorbed by that body and converted back to thermal energy or heat. When thermal radiations falls upon a body, part is absorbed by the body in the form of heat, part is reflected back into space and in some case part can be transmitted through the body.

The basic equation for heat transfer by radiation from a body at temperature T is:

$$q = A\varepsilon\sigma T^4 \qquad (15)$$

where ε is the emissivity of the body. $\varepsilon = 1$ for a perfect black body while real bodies which are gray bodies have an $\varepsilon < 1$.

11.6.15 POROSITY AND PORE SIZE DISTRIBUTION IN A BODY

Porosity refers to volume fraction of void spaces. This void space can be actual space filled with air or space filled with both water and air. Many different definitions of porosity are possible. For non-hygroscopic materials, porosity does not change with change in moisture content. For hygroscopic materials, porosity changes with moisture content. However, such changes during processing are complex due to consideration of bound water and are typically not included in computations.

The distinction between porous and capillary-porous is based on the presence and size of the pores. Porous materials are sometimes defined as those having pore diameter greater than or equal to 10^{-7} m and capillary-porous as one having diameter less than 10^{-7} m. Porous and capillary porous materials were defined as those having a clearly recognizable pore space.

In non-hygroscopic materials, the pore space is filled with liquid if the material is completely saturated and with air if it is completely dry. The amount of physically bound water is negligible. Such a material does not shrink during heating. In non-hygroscopic materials, vapor pressure is a function of temperature only. Examples of non-hygroscopic capillary-porous materials are sand, polymer particles and some ceramics. Transport materials in non-hygroscopic materials do not cause any additional complications as in hygroscopic materials.

In hygroscopic materials, there is large amount of physically bound water and the material often shrinks during heating. In hygroscopic materials there is a level of moisture saturation below which the internal vapor pressure is a function of saturation and temperature. These relationships are called equilibrium moisture isotherms. Above this moisture saturation, the vapor pressure is a function of temperature only and independent of the moisture level. Thus, above certain moisture level, all materials behave non-hygroscopic.

Transport of water in hygroscopic materials can be complex. The unbound water can be in funicular and pendulum states. This bound water is removed by progressive vaporization below the surface of the solid, which is accompanied by diffusion of water vapor through the solid.

Examples of porous materials are to be found in everyday life. Soil, porous or fissured rocks, ceramics, fibrous aggregates, sand filters, snow layers and a piece of sugar or bread are but just a few. All of these materials have properties in common that intuitively lead us to classify them into a single denomination: *porous media.*

Indeed, one recognizes a common feature to all these examples. All are described as "solids" with "holes," i.e., presenting *connected void spaces,* distributed—randomly or quite homogeneously—within a *solid matrix.* Fluid flows can occur within the porous medium, so that we add one essential feature: this *void space* consists of a complex tridimensional network of *interconnected* small empty volumes called *"pores,"* with several continuous paths linking up the porous matrix spatial extension, to enable flow across the sample.

If we consider a porous medium that is not consolidated, it is possible to derive the *particle-size distribution* of the constitutive solid grains. The problem is obvious when dealing with spherical shaped particles, but raises the question of what is meant by particle size in the case of an irregular shaped particle. In both cases, a first intuitive approach is to start with a *sieve analysis.* It consists to sort the constitutive solid particles among various sieves, each one having a calibrated mesh size. The most common type of sieve is a woven cloth of stainless steel or other metal, with wire diameter and tightness of weave controlled to produced roughly rectangular openings of known, uniform size. By shaking adequately the raw granular material, the solid grains are progressively falling through the stacked sieves of decreasing mesh sizes, i.e., a *sieve column.* We finally get separation of the grains as function of their *particle- size* distribution that is also denoted by the porous medium *granulometry.* This method can be implemented for dry granular samples. The *sieve analysis* is a very simple and inexpensive separation method, but the reported granulometry depends very much on the shape of the particles and the duration of the laboratory test, since the sieve will let in theory pass any particle with a smallest cross-section smaller than the nominal mesh opening. For example, one gets very different figure while comparing long thin particles to spherical particles of the same weight.

The definition of a porous medium can be based on the objective of describing flow in porous media. A porous medium is a heterogeneous system consisting of a rigid and stationary solid matrix and fluid filled voids. The solid matrix or phase is always continuous and fully connected. A phase is considered a homogeneous portion of a system, which is separated from other such portions by a definitive boundary, called an interface. The size of the voids or pores is large enough such that the contained fluids can be treated as a continuum. On the other hand, they are small enough that the interface between different fluids is not significantly affected by gravity.

The topology of the solid phase determines if the porous medium is permeable, i.e., if fluid can flow through it, and the geometry determines the resistance to flown and therefore the permeability. The most important influence of the geometry on the permeability is through the interfacial or surface area between the solid phase and the fluid phase. The topology and geometry also determine if a porous medium is isotropic, i.e., all parameters are independent of orientation or anisotropic if the parameters depend on orientation. In multi- phase flow the geometry and surface characteristics of the solid phase determine the fluid distribution in the pores, as does the interaction between the fluids. A porous medium is homogeneous if its average properties are independent of location, and heterogeneous if they depend on location. An example of a porous medium is sand. Sand is an unconsolidated porous medium, and the grains have predominantly point contact. Because of the irregular and angular nature of sand grains, many wedge-like crevices are present. An important quantitative aspect is the surface area of the sand grains exposed to the fluid. It determines the amount of water, which can be held by capillary forces against the action of gravity and influences the degree of permeability.

The fluid phase occupying the voids can be heterogeneous in itself, consisting of any number of miscible or immiscible fluids. If a specific fluid phase is connected, continuous flow is possible. If the specific fluid phase is not connected, it can still have bulk movement in ganglia or drops. For single-phase flow the movement of a Newtonian fluid is described. For two-phase immiscible flow, a viscous Newtonian wetting liquid together with a non-viscous gas are described. In practice these would be water and air.

11.6.16   PORE-SIZE DISTRIBUTION IN POROUS STRUCTURE

A detailed description of the complex tri-dimensional network of pores is obviously impossible to derive. For consolidated porous media, the determination of a *pore-size distribution* is nevertheless useful. For those particular media, it is indeed impossible to handle any particle-size distribution analysis.

One approach to define a *pore size* is in the following way: the *pore diameter* δ at a given point within the pore space is the diameter of the largest sphere that contains this point, while still remaining entirely within the pore space. To each point of the *pore space* such a "diameter" can be attached rigorously, and the *pore-size distribution* can be derived by introducing the *pore-size density function* $\theta(\delta)$ defined as the fraction of the total void space that has a pore diameter comprised between δ and $\delta + d\delta$. This distribution is normalized by the Eq. (16) of Appendix:

$$\int_0^\infty \theta(\delta)d\delta = 1 \qquad (16)$$

A porous structure should be:
- A material medium made of heterogeneous or multiphase matter. At least one of the considered phases is not solid. The solid phase is usually called the *solid matrix*. The space within the porous medium domain that is not part of the *solid matrix* is named *void space* or *pore space*. It is filled by gaseous and/or liquid phases.
- The solid phase should be distributed throughout the porous medium to draw a network of pores, whose characteristic size can vary great-

ly. Some of the pores comprising the void space must enable the flow across the solid matrix, so that they should then be interconnected.

- The interconnected pore space is often denoted as the *effective pore space*, while unconnected pores may be considered from the hydrodynamic point of view as part of the solid matrix, since those pores are ineffective as far as flow through the porous medium is concerned. They are *dead-end pores* or *blind pores*, that contain stagnant fluid and no flow occurs through them.

A porous material is a set of pores embedded in a matrix of mostly solid material. The pores are the voids in the material itself. Pores can be isolated or interconnected. Furthermore, a pore can contain a fluid or a vapor, but it can also be empty. If the pore is completely filled with the fluid, it will be called saturated and if it is partially filled, it will be called non-saturated. So the porous material is primarily characterized by the content of its voids and not by the properties of the material itself. Fig.1 of Appendix gives a sketch of a porous material.

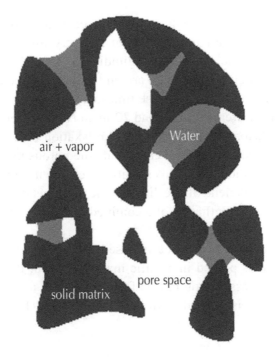

FIGURE 1   A 2D sketch of a non-saturated porous material.

If the pores are not interconnected very well, the relaxation-time distribution of an NMR (Nuclear Magnetic Resonance) spin-echo measurement can be interpreted in terms of a pore-size distribution (PSD). For magnetically doped materials like clay and red-clay this so-called relaxometry technique gives a pore-size distribution between 100 nm and 100 μm, which is also the range of the majority of the pores in these materials. NMR (Nuclear Magnetic Resonance) can be used for spectroscopy, because different nuclei resonate at different frequencies and can therefore be distinguished from each other. Not only nuclei, but also different isotopes can be distinguished. Since also the surrounding of the nucleus has an effect on the exact resonance frequency, NMR spectroscopy is also used to distinguish specific molecules. By manipulating the spatial dependence of the magnetic field strength and the frequency of the RF excitation, the NMR sensitive region can be varied. This enables a noninvasive measurement of the spatial distribution of a certain nucleus and is called NMR Imaging (MRI).

In many NMR experiments it was noticed that liquids confined in porous materials exhibit properties that are very different from those of the bulk fluid. The so-called longitudinal ($T1$) and transverse ($T2$) relaxation time of bulk water, e.g., are on the order of seconds, whereas for water in a porous material these times can be on the order of milliseconds. The measurement of $T1$ and $T2$ in an NMR experiment is often called NMR relaxometry. The transverse relaxation time is more sensitive to local magnetic field gradients inside the porous material than the longitudinal relaxation time. This sensitivity can be used to measure the self-diffusion coefficient of the liquid. The interpretation of the measured self-diffusion coefficient of a confined liquid is often called NMR diffusometry.

Nuclear magnetic resonance is based on the following principle. When a nucleus is placed in a static magnetic field, the nuclear spin $\underline{I}$ will start to presses around this field, since the magnetic moment $\bar{\mu}$ of the nucleus is related to the nuclear spin $\bar{I}$ (Fig. 2 of Appendix).

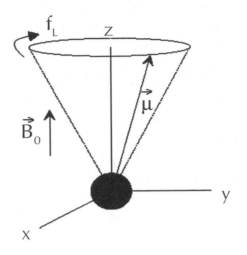

FIGURE 2   Larmor precession of a nuclear magnetic moment in a magnetic field.

The frequency of this precession motion is called the Larmor frequency:

$$f_L = \frac{\gamma}{2\pi} B_0 \tag{17}$$

where B_0 is the magnitude of the static magnetic field, which is usually taken aligned with the z-axis, f_L is the Larmor frequency and γ is the gyromagnetic ratio of the nucleus.

The NMR resonance condition, Eq. (17) of Appendix, states that the Larmor frequency depends linearly on the magnetic field. Normally one starts to assume that the magnetic field in the porous material is equal to the magnetic field generated by the experimental setup. This can be either the magnetic field emerging from a permanent magnet, an electromagnet, or a superconducting magnet. Frequently, an extra magnetic field gradient is added to the main magnetic field. This magnetic field gradient is used to discriminate spins at a certain position from spins at other positions. It is the basic principle of NMR Imaging (MRI). However, the magnetic field inside the porous sample can deviate largely from the magnetic field applied externally.

Because the magnetic susceptibility of the porous material differs from that of the surrounding air, the magnetic field inside the porous sample will deviate from the magnetic field that is present in the sample chamber or insert. Apart from this, the magnetic field in the pores of the material may differ from that in the bulk matrix. Consider two media with a different susceptibility. If the magnetic susceptibility of the sphere is larger (Fig. 3 of Appendix on the left) than that of the environment, the magnetic field inside this sphere is larger than the external magnetic field and the sphere is called paramagnetic. If, on the other hand, the susceptibility of the sphere is smaller (Fig. 3 of Appendix on the right) than that of the environment, the magnetic field inside the sphere is smaller than the external magnetic field and the sphere is called diamagnetic.

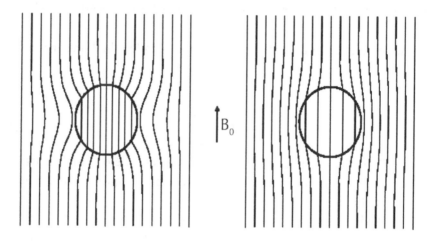

FIGURE 3 Disturbance of homogeneous magnetic field B_0 by an object with different susceptibility. Plotted are the magnetic field lines. On the left: a paramagnetic sphere; on the right: a diamagnetic sphere.

The amount of water in a porous body such as the textiles at the EMC is defined as bound water and it is absorbed by the textile fibers. When the textile is unable to absorb more water, all excess water is defined as unbound moisture. The unbound moisture is often found as a continuous liquid within the porous material.

Drying of porous media is accomplished by vaporizing the water and to do this the latent heat of vaporization must be supplied. There are, thus, two important process-controlling factors that enter into the process of drying:
- transfer of heat to provide the necessary latent heat of vaporization,
- movement of water or water vapor through textiles and then away from it to effect separation of water.

11.6.17 BASIC FLOW RELATIONS IN POROUS BODY

The motion of a fluid is described by the basic hydrodynamic equations, the continuity equation:

$$\partial_t \rho + \nabla.(\rho u) = 0 \qquad (18)$$

which expresses the conservation of mass, and the momentum:

$$\partial_t (\rho u) + \nabla.(\rho u) = \nabla\ p + \nabla.\tau + \rho g \qquad (19)$$

which expresses the conservation of momentum. Here ρ is the fluid density, u the fluid velocity, p the hydrostatic pressure, τ the fluid stress tensor, and g the acceleration due to external forces including e.g. the effect of gravity on the fluid.

The Eq. (20) of Appendix for energy conservation can be written as:

$$\rho\frac{d\hat{u}}{dt} + p(\nabla.u) = \nabla.(k\nabla T) + \Phi \qquad (20)$$

where T is temperature, k the coefficient of thermal conductivity of the fluid, Φ the viscous dissipation function, and the density of thermal energy $\hat{u} = \hat{u}(p,T)$ is often approximated such that $d\hat{u} \approx c_v dT$, where c_v is the specific heat.

At low Reynolds numbers, the most important relation describing fluid transport through porous media is Darcy's law.

$$q = -\frac{k}{\mu}\nabla p \qquad\qquad (21)$$

where q is the volumetric fluid flow through the (homogeneous) medium and k is the permeability coefficient that measures the conductivity to fluid flow of the porous material.

11.6.18 TRANSPORT MECHANISMS IN POROUS MEDIA

The study of flow systems, which compose, of a porous medium and a homogenous fluid has attracted much attention since they occur in a wide range of the industrial and environmental applications. Examples of practical applications are: flow past porous scaffolds in bioreactors, drying process, electronic cooling, ceramic processing, and overland flow during rainfall, and ground-water pollution.

In the single-domain approach, the composite region is considered as a continuum and one set of general governing equations is applied for the whole domain. The explicit formulation of boundary conditions is avoided at the interface and the transitions of the properties between the fluid and porous medium are achieved by certain artifacts. Although this method is relatively easier to implement, the flow behavior at the interface may not be simulated properly, depending on how the code is structured.

In the two-domain approach, two sets of governing equations are applied to describe the flow in the two regions and additional boundary conditions are applied at the interface to close the two set of equations. This method is more reliable since it tries to simulate the flow behavior at the interface. Hence, in the present study, the two-domain approach, and the implementation of the interface boundary conditions, will be considered.

Fluid flow in a porous medium is a common phenomenon in nature, and in many fields of science and engineering. Important everyday flow phenomena include transport of water in living plants and trees, and fertilizers or wastes in soil. Moreover, there is a wide variety of technical processes that involve fluid dynamics in various branches of process industry. The importance of improving our understanding of such processes arises from the high amount of energy consumed by them. In oil recovery, for example, a typical problem is the amount of un-recovered oil left in oil reservoirs by traditional recovery techniques. In many cases the porous structure of the medium and the related fluid flow are very complex, and detailed studies of these flows pose demanding tasks even in the case of stationary single-fluid flow. In experimental and theoretical work on fluid flow in porous materials it is typically relevant to find correlations between material characteristics, such as porosity and specific surface area, and flow properties. The most important phenomenological law governing the flow properties, first discovered by Darcy, defines the permeability as conductivity to fluid flow of the porous material. Permeability is given by the coefficient of linear response of the fluid to a non-zero pressure gradient in terms of the flux induced.

Some of the material properties that affect the permeability, e.g. tortuosity, are difficult to determine accurately with experimental techniques, which have been, for a long time, the only practical way to study many fluid-dynamical problems. Improvement of computers and the subsequent development of methods of computational fluid dynamics (CFD) have gradually made it possible to directly solve many complex fluid-dynamical problems. Flow is determined by its velocity and pressure fields, and the CFD methods typically solve these in a discrete computational grid generated in the fluid phases of the system. Traditionally CFD has concentrated on finding solution to differential continuum equations that govern the fluid flow. The results of many conventional methods are sensitive to grid generation, which most often can be the main effort in the application. A successfully generated grid is typically an irregular mesh including knotty details that follow the expected streamlines.

Transport in a porous media can be due to several different mechanisms. Three of these mechanisms are often considered most dominant: molecular diffusion, capillary diffusion, and convection (*Darcy flow*).

The *Darcy law* has been derived as follows: we consider a macroscopic porous medium which has a cross section A and overall length L, and we impose an oriented fluid flow rate $\vec{Q}$, to flow through it .When a steady state is reached, the induced hydrostatic pressure gradient $\vec{\nabla}p$ is related to $\vec{Q}$ by the vectorial formula:

Fluid dynamics (also called fluid mechanics) is the study of moving (deformable) matter, and includes liquids and gases, plasmas and, to some extent, plastic solids. From a 'fluid-mechanical' point of view, matter can, in a broad sense, be considered to consist of fluid and solid, in a one-fluid system the difference between these two states being that a solid can resist shear stress by a static deformation, but a fluid can not . Notice also that thermodynamically a distinction between the gas and liquid states of matter cannot be made if temperature is above that of the so-called critical point, and below that temperature the only essential differences between these two phases are their differing equilibrium densities and compressibility.

$$\frac{\vec{Q}}{A} = \frac{\overline{K}}{\mu_f} \cdot \left(\vec{\nabla}p - \rho_f \cdot \vec{g}\right) \Leftrightarrow \vec{v}_m = \frac{\overline{K}}{\mu_f} \cdot \left(\frac{\Delta p}{L} - \rho_f \cdot \vec{g}\right) \qquad (22)$$

where $\vec{g}$ is the acceleration of the gravity field, ρ_f and μ_f are respectively the *specific mass* and the *dynamic viscosity* of fluid, $\vec{v}_m$ the *filtration velocity* over the cross section A. Eq. (22) of Appendix defines a second order symmetrical tensor $\overline{K}$, the *permeability*. It takes into account the macroscopic influence of the porous structure from the "resistance to the flow" point of view. The more permeable a porous medium is, the less it will resist to an imposed flow. The *permeability* is an intrinsic property of the porous matrix, based only on geometrical considerations, and is expressed in $[m^2]$. The tensorial character of $\overline{K}$ reflects the porous matrix *anisotropy*.

At the surface of the textile, two processes occur simultaneously in drying: heat transfer from the air to the drying surface and mass transfer from the drying surface to the surrounding air. The energy transfer between a surface and a fluid moving over the surface is traditionally described by convection. The unbound moisture on the surface of the material is first vaporized during the constant drying rate period.

Heat transfer by convection is described as:

$$\frac{dQ}{dt} \overline{h}A\left(T_A - T_S\right) \qquad (23)$$

where dQ/dt is the rate of heat transfer, $h [W/m^2 K]$ is the average heat transfer coefficient for the entire surface, A. T_S is the temperature of the material surface and T_A is the air temperature. The temperature on the surface is close to the wet bulb temperature of the air when unbound water is evaporated.

A similar equation describes the convective mass transfer. The total molar transfer rate of water vapor from a surface, dN_v / dt [kmol/s], is determined by:

$$\frac{dN_v}{dt} = \bar{h}_m A \left(C_{v,A} - C_{v,S} \right) \tag{24}$$

where $\bar{h}_m$ (m/s) is the average convection mass transfer coefficient for the entire surface, $C_{v,A}$ is the molar concentration of water vapor in the surrounding air and $C_{v,S}$ is the molar concentration on the surface of the solid with the units of (kmol/m³). During the constant drying rate period the drying rate is controlled by the heat and/or mass transfer coefficients, the area exposed to the drying medium, and the difference in temperature and relative humidity between the drying air and the wet surface of the material.

The average convection coefficients depend on the surface geometry of the material and the flow conditions. The heat transfer coefficient, h, can be determined by the average Nusselt number, $\overline{Nu}$:

$$\overline{Nu} = \frac{\bar{h} L}{k_A} = f \left(\text{Re}, \text{Pr} \right) \tag{25}$$

where k_A is the heat conductivity for the air and L is the characteristic length of the surface of interest. $\overline{Nu}$ shows the ratio of the heat transfer that depends on convection to the heat transfer that depends on conduction in the boundary layer. The Nusselt number is a function of the Reynold number, Re, and the Prantdl-number, Pr. Pr is the relation between the thickness of the thermal and the velocity boundary layers. If $Pr = 1$, the thickness of the thermal and velocity boundary layers are equal. For air $Pr = 0.7$. To determine the mass transfer coefficient, $\bar{h}_m$, the average Sherwood-number, $\overline{Sh}$ is used.

$$\overline{Sh} = \frac{\overline{h}_m L}{D_{AS}} = f\left(\mathrm{Re}, Sc\right) \qquad (26)$$

where D_{AS} is the diffusion coefficient. $\overline{Sh}$ is a function of the Reynold number, Re, and the Schmidt number, Sc, which is the relation between the thickness of the concentration and the velocity boundary layers.

Water vapor in the porous media can move by molecular or Fickian diffusion if the pores are large enough. Molecular diffusion is described by Fick's law,

$$J = -D\frac{\partial c}{\partial x} \qquad (27)$$

where, D is the molecular diffusivity.

In a fiber such as textile the diffusion does not only depend on the difference in concentration but also on the characteristics of the textile. He describes the moisture movement as being dependent on the density of the solid, which is a function of the moisture content as the fiber swells or shrinks in response to the moisture that is present.

Flow in porous media plays an important role in many areas of science and engineering. Examples of the application of porous media flow phenomena are as diverse as flow in human lungs or flow due to solidification in the mushy zone of liquid metals.

Flow in porous media is difficult to be accurately modeled quantitatively. Richards equation can give good results, but needs constitutive relations. These are usually empirically based and require extensive calibration. The parameters needed in the calibration are amongst others: capillary pressure and pressure gradient, volumetric flow, liquid content, irreducible liquid content, and temperature. In practice it is usually too demanding to measure all these parameters.

The description of the behavior of fluids in porous media is based on knowledge gained in studying these fluids in pure form. Flow and transport phenomena are described analogous to the movement of pure fluids without the presence of a porous medium. The presence of a permeable solid influences these phenomena significantly. The individual description of the movement of the fluid phases and their interaction with the solid phase is modeled by an up-scaled porous media flow equation. The concept of up-scaling from small to large scales is widely used in physics. Statistical physics translates the description of individual molecules into a continuum description of different phases, which in turn is translated by volume averaging into a continuum porous medium description [2–4, 5–18, 20, 22–46, 48–63, 65–95].

KEYWORDS

- **Black body**
- **Breathable**
- **Bulk heating**
- **Convective heat transfer coefficient**
- **Dynamic viscosity**
- **Effective thermal conductivity**
- **Emissivity**
- **Filtration velocity**
- **Granulometry**
- **Heat transfer**
- **Hydrophobic fabric**
- **Hygroscopic fibers**
- **Lower body temperature**
- **Miracle fibers**
- **Permeability**
- **Pore-size distribution**
- **Sieve column**
- **Solid matrix**
- **Water vapor resistance**
- **Wet fabrics**

REFERENCES

1. Armour, J.; Cannon, J. Fluid Flow Through Woven Screens, *AIChE J.* **1968**, *14(3)*, 415–420.
2. ASTM D737–75, Standard Test Methods for Air Permeability of Textile Fabrics.
3. ASTM 96–95, Water Vapor Transmission of Materials.
4. Arnold, G. and Fohr, J. P. Slow Drying Simulation in Thick Layears of Granular Products, *Int. J. Heat Mass Transfer* **1988**, *31*(12), 2517–2562.
5. Azizi, S.; Moyne, C.; and Degiovanni, A. Approche Experimentale et theorique de la Conductivite Thermique des Milieux Poreux Humides, *Int. J. Heat Mass Transfer* **1988**, *31(11)*, 2305–2317.
6. Backer, S. The Relationship Between the Structural Geometry of a Textile Fabric and its Physical Properties, Part IV: Interstice Geometry and Air Permeability, *Textile Res. J.* **1951**, *21*, 703–714.
7. Barnes, J. and Holcombe, B. Moisture Sorption and Transport in Clothing During Wear, *Textile Res. J.* **1996**, *66* (*12*), 777–786.
8. Bartles, V. T. Survey on the Moisture Transport Properties of Foul Weather Protective Textiles at Temperatures Around and Below the Freezing Point, Technical Report (*11674*), Hohenstein Institute of Clothing Physiology, Boennigheim, Germany, **2001**.
9. Bears, J. *"Dynamics of Fluids in Porous Media,"* Elsevier, New York, **1972**.
10. Black, W. Z. and Hartley, J. G. *Thermodynamics,* Harper & Row, New York, **1985**.
11. BS 4407, Quantitative Analysis of Fiber Mixtures, **1997**.
12. BS 7209, Specification for Water Vapor Permeable Apparel Fabrics, **1990**.
13. CAN2–4.2-M77, Method of Test for Resistance of Materials to Water Vapor Diffusion (Control Dish Method), **1977**.
14. CGSB-4.2 (49)-M91, Resistance of Materials to Water Vapor Diffusion.
15. Chen, C. S.; Johnson, W. H. Kinetics of Moisture Movement in Hygroscopic Materials, In: Theoretical Considerations of Drying Phenomenon, *Trans. ASAE.,* **1969**, *12*, 109–113.
16. Chen, P.; Pei, D. A Mathematical Model of Drying Process, *Int. J. Heat Mass Transfer* **1988**, *31(12)*, 2517–2562.
17. Chen, P.; Schmidt, P. S. An Integral Model for Drying of Hygroscopic and Non-hygroscopic Materials with Dielectric Heating, Drying Technol. **1990**, *8(5),* 907–930.
18. Chen, P.; Schmidt, P. S. A Model for Drying of Flow-through Beds of Granular Products with Dielectric Heating, in: *Transport Phenomena in Materials Processing, American Society of Mechanical Engineers, Heat Transfer Division,* (Publication) HDT, *146*, ASME, New York, **1990**, 121–127.
19. Davis, A.; James, D. Slow Flow Through a Model Fibrous Porous Medium, *Int. J. Multiphase Flow 22*, 969–989 **1996**,.
20. Dietl, C.; George, O. P.; Bansal, N. K. Modeling of Diffusion in Capillary Porous Materials During the Drying Process, Drying Technol. **1995**, 13(1&2), 267–293.
21. Ea, J. Y. Water Vapor Transfer in Breathable Fabrics for Clothing, PhD thesis, University of Leeds, **1988**.
22. Flory, P. J. Statistical Mechanics of Chain Molecules, Interscience Pub. NY, **1969**.

23. Francis, N. D.; Wepfer, W. J. Jet Impeingement Drying of a Moist Porous Solid, *Int. J. Heat Mass Transfer* **1996**, *39(9)*, 1911–1923.
24. Gerald, C. F.; Wheatley, Applied Numerical Analysis, Fourth ed., Addison-Weseley, Reading, MA, **1989**.
25. Ghali, K.; Jones, B.; Tracy, E. Experimental Techniques for Measuring Parameters Describing Wetting and Wicking in Fabrics, *Textile Res. J.,* **1994**, 106–111.
26. Gibson, P.; Elsaiid, A.; Kendrick, C. E.; Rivin, D.; Charmchi, M. A Test Method to determine the relative Humidity Dependence of the Air Permeability of Textile Materials, *J. Testing Eval.* **1997**, *25(4)*, 416–423.
27. Gibson, P.; Charmchi, M. The Use of Volume-Averaging Techniques to Predict Temperature Transients Due to Water Vapor Sorption in Hygroscopic Porous Polymer Materials, *J. Appl. Polym. Sci.* **1997**, *64*, 493–505.
28. Ghali, K.; Jones, B.; Tracy, E. Modeling Heat and Mass Transfer in Fabrics, *Int. J. Heat Mass Transfer* **1995**, *38(1)*, 13–21.
29. Gennes, P. G. Scaling Concepts in Polymer Physics, 3ʳᵈ ed., Cornell University Press, Ithaca, NY, **1988**.
30. Givoni, B. and Goldman, R. F. Predicting Metabolic Energy Cost, *J. Appl. Physiol.* **1971**, *30(3)*, 429–433.
31. Green, J. H. "An Introduction to Human Physiology" J. Comyn, Ed., Elsevier Applied Science Publishers, London, **1985**.
32. Greenkorn, R. A. "Flow Phenomena in Porous Media" Marcel Dekker, New York, **1984.**
33. Hadley, G. R. Numerical Modeling of the Drying of Porous Materials, in: *Proceedings of The Fourth International Drying Symposium,* **1984**, *1*, 151–158.
34. Haghi, A. K. Moisture permeation of clothing, *JTAC* **2004**, *76*, 1035–1055.
35. Haghi, A. K. Thermal analysis of drying process, *JTAC* **2003**, *74*, 827–842.
36. Haghi, A. K. Some Aspects of Microwave Drying, The Annals of Stefan cel Mare University, Year VII, **2000**, *14*, 22–25.
37. Haghi, A. K. A Thermal Imaging Technique for Measuring Transient Temperature Field- An Experimental Approach, The Annals of Stefan cel Mare University, Year VI, **2000**, *12*, 73–76.
38. Haghi, A. K. Experimental Investigations on Drying of Porous Media using Infrared Radiation, *Acta Polytechnica,* **2001**, *41(1)*, 55–57.
39. Haghi, A. K. A Mathematical Model of the Drying Process, *Acta Polytechnica* **2001**, *41(3)* 20–23.
40. Haghi, A. K. Simultaneous Moisture and Heat Transfer in Porous System, *Journal of Computational and Applied Mechanics* **2001**, *2(2)*, 195–204.
41. Haghi, A. K. A Detailed Study on Moisture Sorption of Hygroscopic Fiber, *Journal of Theoretical and Applied Mechanics* **2002**, *32(2)* 47–62.
42. Haghi, A. K. A mathematical Approach for Evaluation of Surface Topography Parameters, *Acta Polytechnica* **2002**, *42(2)*, 35– 40.
43. Haghi, A. K. Mechanism of Heat and Mass Transfer in Moist Porous Materials, *Journal of Technology,* **2002**, *35(F)* 1–16.
44. Haghi, A. K. A Study of Drying Process, *H.J.I.C.,* **2002**, *30*, 261–269.
45. Haghi, A. K. Experimental Evaluation of the Microwave Drying of Natural Silk *J. of Theoretical and Applied Mechanics* **2003**, *33*, 83–94.

46. Haghi, A. K.; Mahfouzi, K.; Mohammadi, K. The effects of Microwave Irradiations on Natural Silk, *JUCTM* **2002**, *38*, 85–96.

47. Haghi, A. K. The Diffusion of Heat and Moisture Through Textiles, *International Journal of Applied Mechanics and Engineering,* **2003**, *8(2)*, 233–243.

48. Haghi, A. K.; Rondot, D. Heat and Mass Transfer of Leather in The Drying Process, *IJC&Chem. Engng.* **2004**, *23*, 25–34.

49. Haghi, A. K.; Heat and Mass Transport Through Moist Porous Materials, *14th Int. Symp. on Transport Phenomena Proc.,* 209–214, 6–9 July **2003**, Indonesia.

50. Hartley, J. G. Coupled Heat and Moisture Transfer in Soils: A Review, Adv. Drying *4,* 199–248 (199–248).

51. Higdon, J.; anf Ford, G. Permeability of Three-Dimensional Models of Fibrous Porous Media, *J. Fluid Mechan,* **1996**, *308*, 341–361.

52. Hong, K.; Hollies, N. R. S.; Spivak, S. M. Dynamic Moisture Vapor Transfer Through Textiles, Part I: Clothing Hygrometry and the Influence of Fiber Type, *Textile Res. J.* **1988**, *58(12)*, 697–706.

53. Hsieh, Y. L.; Yu, B.; Hartzell, M. Liquid Wetting Transport and Retention Properties of Fibrous Assemblies, Part II: Water Wetting and Retention of 100% and Blended Woven Fabrics, *Textile Res. J.* **1992**, *62(12)*, 697–704.

54. Huh, C.; Scriven, L. E. Hydrodynamic Model of Steady Movement of a Solid-Liquid-Fluid Contact Line, *J. Coll. Inter. Sci.,* **1971**, *35*, 85–101.

55. Incropera, F. P.; Dewitt, D. P. Fundamentals of Heat and Mass Transfer, second ed., Wiley, New York, **1985**.

56. ISO 11092, Measurement of Thermal and Water-vapor Resistance under Steady-state Conditions (Sweating Guarded-hotplate Test), **1993**.

57. Ito, H. and Muraoka, Y. Water Transport Along Textile Fibers as Measured by an Electrical Capacitance Technique, *Textile Res. J.* **1993**, *63(7)*, 414–420.

58. Jackson, J.; James, D. The Permeability of Fibrous Porous Media, *Can. J. Chem. Eng.* **1986**, *64*, 364–374.

59. Jacquin, C. H.; Legait, B. Influence of Capillarity and Viscosity During Spontaneous Imbibition in Porous Media and Capillaries, *Phy. Chem. Hydro.* **1984**, *5*, 307–319.

60. Jirsak, O.; Gok, T.; Ozipek, B.; Pau, N. Comparing Dynamic and Static Methods for Measuring Thermal Conductive Properties of Textiles, *Textile Res. J.* **1998**, *68(1)*, 47–56.

61. Kaviany, M. "Principle of Heat Transfer in Porous Media," Springer, New York, **1991**.

62. Keey, R. B. "Drying: Principles and Practice," Oxford, Pergamon, **1975**.

63. Keey, R. B. "Introduction to Industrial Drying Operations," Oxford, Pergamon, **1978**.

64. Keey, R. B. The Drying of Textiles, *Rev. Prog. Coloration* **1993**, *23*, 57–72.

65. Kulichenko, A.; Langenhove, L. The Resistance to Flow Transmission of Porous Materials, *J. Textile Inst.* **1992**, *83(1)*, 127–132.

66. Kyan, C.; Wasan, D.; Kintner, R. Flow of Single-Phase Fluid through Fibrous Beds, *Ind. Eng. Chem. Fundament.* **1970**, *9(4)*, 596–603.

67. Le, C. V.; Ly, N. G. Heat and Moisture Transfer in Textile Assemblies, Part I: Steaming of Wool, Cotton, Nylon, and Polyester Fabric Beds, *Textile Res. J.* **1995**, *65(4)*, 203–212.

68. Le, C. V.; Tester, D. H.; Buckenham, P. Heat and Moisture Transfer In Textile Assemblies, Part II: Steaming of Blended and Interleaved Fabric-Wrapper Assemblies, *Textile Res. J.* **1995**, *65(5)*, 265–272.

69. Lee, H. S.; Carr, W. W.; Becckham, H. W.; Wepfer, W. J. Factors Influencing the Air Flow Through Unbacked Tufted Carpet, *Textile Res. J.* **2000**, *70*, 876–885.

70. Lee, H. S. Study of the Industrial Through-Air Drying Process For Tufted Carpet, Doctoral thesis, Georgia Institute of Technology, Atalnta, GA, **2000**.

71. Luikov, A. V. Heat and Mass Transfer in Capillary Porous Bodies, Pergamon Press, Oxford, **1966**.

72. Luikov, A. V. Systems of Differential Equations of Heat and Mass Transfer in Capillary Porous Bodies, *Int. J. Heat Mass Transfer 18*, 1–14, **1975**.

73. Metrax, A. C. and Meredith R. J. Industrial Microwave Heating, Peter Peregrinus Ltd, London, England, **1983**.

74. Mitchell, D. R.; Tao, Y. and Besant, R. W. Validation of Numerical Prediction for Heat Transfer with Airflow and Frosting in Fibrous Insulation, paper 94–WA/HT-10 in *"Proc. ASME Int. Mech. Eng. Cong.,* Chicago," Nov. **1994**.

75. MOD Specification UK/SC/4778A SCRDE, Moisture Vapor Transmission Test Method.

76. Morton, W. E. and Hearle, J. W. S. Physical Properties of Textile Fibers" 3rd ed., Textile Institute, Manchester, U.K., **1993**.

77. Moyene, C.; Batsale, J. C.; Degiovanni, A. Approche Experimentale et theorique de la Conductivite Thermique des Milieux Poreux Humides, II: Theorie, *Int. J. Heat Mass Transfer* **1988**, *31(11)*, 2319–2330.

78. Mujumdar, A. S. Handbook of Industrial Drying, Marcel Decker, New York, **1985**.

79. Nasrallah, S. B. and Pere, P. Detailed Study of a Model of Heat and Mass Transfer During Convective Drying of Porous Media, *Int. J. Heat Mass Transfer* **1988**, *31(5)*, 957–967.

80. Nossar, M. S.; Chaikin, M.; Datyner, A. High Intensity Drying of Textile Fibers, Part I: The Nature of The Flow of Air Through Beds of Drying Fibers, *J. Textile Inst,* **1973**, *64*, 594–600.

81. Patankar, S. V. "Numerical Heat Transfer and Fluid Flow," Hemisphere Publishing, NY, **1980**.

82. Penner, S. and Robertson, A. Flow through Fabric-Like Structures, *Textile Res. J. 21*, 775–788 **1951**,.

83. Provornyi, S.; Slobodov, E. Hydrodynamics of a Porous Medium with Intricate Geometry, *Theoret. Foundat. Chem. Eng.* **1995**, *29(1)*, 1–5.

84. Rainard, L. W. Air Permeability of Fabrics I, *Textile Res. J.* **1946**, *16*, 473–480.

85. Rainard, L. W. Air Permeability of Fabrics II, *Textile Res. J.* **1947**, *17*, 167–170.

86. Renbourne, E. T. "Physiology and Hygiene of Materials and Clothing," Merrow Publishing Company, Herts, England, **1971**.

87. Saltiel, C. and Datta, A. S. Heat and Mass Transfer in Microwave Processing, *Advances in Heat and Mass Transfer* **1999**, *33*, 1–94.

88. Sanga, E.; Mujumdar, A. S.; Raghavan, G. S. Heat and Mass Transfer in Non-homogeneous Materials under Microwave Field, Presented at The 50th Canadian Chemical Engineering Conference, Montreal, Canada, October 15–18, **2000**.

89. Simacek, P.; Advani, S. Permeability Model for a Woven Fabric, *Polym. Compos.* **1996**, *17(6)*, 887–899.
90. Spencer-Smith, J. L. The physical Basis of Clothing Comfort, Part 3: Water Vapor Transfer Through Dry Clothing Assemblies, *Clothing Res. J.* **1977**, *5*, 82–100.
91. Spilman, L.; Goren, S. Model for Predicting Pressure Drop and Filtration Efficiency in Fibrous Media, *Environ. Sci. Techno.* **1968**, *2*, 279–287.
92. Stanish, M. A.; Schajer, G. S.; Kayihan, F. A Mathematical Model of Drying for Hygroscopic Porous Media, *AIChE J.* **1986**, *32(8)*, 1301–1311.
93. Toei, R. Drying Mechanism of Capillary Porous Bodies, *Adv. Drying* **1983**, *2*, 269–297.
94. Umbach, K. H. Investigation of Constructional Principles for Clothing Textiles Made of Synthetic Fibers Worn Next to the Skin with Good Comfort Properties, Technical Report no. AiF 3653, Hohenstein Institute of Clothing Physiology, Boennigheim, Germany, **1977**.
95. Umbach, K. H. Moisture Transport and Wear Comfort in Microfiber Fabrics, *Melliand Engl,* **1993**, *74*, E78–E80.
96. Von Hippel, A. R. "Dielectric Materials and Applications," MIT Press, Boston, **1954**.
97. Waananen, K. M.; Litchfield, J. B.; Okos, M. R. Classification of Drying Models foe Porous Solids, *Drying Technol.* **1993**, *11(1)*, 1–40.
98. Watkins, D. A.; Slater, K. The Moisture Vapor Permeability of Textile Fabrics, *J. Textile Inst.* **1981**, *72*, 11–18.
99. Williams, A. "Industrial Drying," Gardner/Leonard Hill Publishers, London, England, **1971**.

CHAPTER 12

NANOPATTERNED IMPLANTS LOADED WITH DRUGS

A. L. IORDANSKII, S. Z. ROGOVINA, I. AFANASOV, and
A. A. BERLIN

CONTENTS

12.1   INTRODUCTION

Challenges and development perspectives on nanopatterned implants loaded with drugs intended to replace or improve human organs and tissues are analyzed. An innovative approach to polymer and composite design combines the surface modification on the molecular and nanosized levels, formation of the implant matrix as hybrid nanocomposite, and drug encapsulation aimed at ensuring their targeted and programmed delivery. The economic and scientometric situations in the world development of composite and hybrid implant systems are briefly described. The major fields in nanoimplantology for the nearest decade are represented, including cardiology, ophthalmology, genitourology and orthopedy. The prospects for biodegradable polymers, such as poly(α-hydroxy acids) (PLA PGA, PLGA) and poly(β-hydroxyalkanoates) (PHB, PHVB), are considered, as well as nanoscale biochips and sensors (NEMS), miniature electromechanical systems (MEMS), and neurological conduits.

When designing the new generation of medical implants, it should be considered many factors and processes that guarantee biomedical efficiency, the retention of the necessary properties during the service life, patient's comfort, as well as other parameters characterizing implant behavior in biological media [1–5]. In addition to the above requirements, the implants should also sustain drug delivery [6–10] and be arranged with nanoscale structures [11, 12]. The latter requirement is determined by implant - cell interactions at the receptor and immune levels, which are most effective if composite implants have a nanostructured surface [13, 14] that will be shown below in the sections of this review.

The implanted medical device (IMD) is an intracorporeal system (device or material) intended for (a) the replacement of lost organs, tissues, or cell assembly; (b) an enhance in the efficiency of the body's biological function; and (c) a patient post-surgery rehabilitation and mechanical support at arrangement in vivo. Implants differ from transplants in their artificial origin, i.e., implants are manufactured under plant, pilot, or laboratory conditions. Traditionally, implants are produced from metals, inorganic compounds, and polymers. Recently, biocomposites, multilayered and hybrid constructs, and nanoscale modifications of implants have been elaborated. Among the IMD, specific features are their biocompatibility and, when necessary, controlled biodegradability. State-of-the-art implants include electronic or computer devices (pacemakers and auricular prostheses), trigger regulators, and feedback systems, preferably wireless. A separate group of implants is the combined systems for dosed and targeted drug delivery in the body. Unlike traditional drug formulation, the active substance in this case is encapsulated directly into the matrix or reservoir of an implanted device as, e.g., in the combined cardiac stents with an active polymer coating. This group of implants is appropriate for therapeutic procedures, surgery (injectors), and diagnosing (in vivo electronic chips).

Note that, if both passive and active transports may be implemented in the case of the nanoparticles assisted drug delivery, the drug encapsulated in the implant enters the adjacent tissues exclusively via a combination of the hydrolytic (enzymatic) process [15] and diffusion [16]. This suggests that, in order to provide the appropriate drug flux from the implant's matrix that is necessary for therapy, the optimally high sorption capacity of implant is required, i.e., the critical concentration of the functional groups in polymer for drug immobilization.

To ensure the efficient function and safety of the implants and coordination of the efforts on their design and commercial promotion, European Union has issued a set of directives (for example, [17] that strictly defines what is an IMD. Over the last 3 years, Russian metrology has also formulated a number of requirements to IMDs. Analogous directives have also been elaborated by the corresponding authorities in the United States [18] and Canada [19].

In 2010, the manufacture of biomedical devices and functional implants worldwide reached $272.2 billion at market price (Table 1). According to the expert forecasts, by the year 2015, this value will increase by 65%, despite the economic crisis and will almost double in Eastern European countries (from $4.1 billion in 2010 to $7.8 billion in 2015). The United States, which occupies the widest sector on the market of medical products and devices (over 40%), is first in the list of manufacturers of implanted systems, followed by EU countries (Western European countries, 27%) [20].

TABLE 1   Dynamics and forecast for the world market of implanted medical devices and prostheses (billions of $$) [18].

Years	2004	2006	2008	2010	2015(forecast)
United State and Canada	83.14	93.97	105.88	118.23	164.63
Western Europe	52.50	59.04	65.65	73.11	103.00
Japan	24.14	26.64	29.31	32.65	44.76
Asian Countries (with out Japan)	15.72	18.36	23.13	30.20	54.97
South America	3.48	3.94	4.83	6.11	11.18
Eastern Europe	3.98	4.50	5.96	1.73	14.90
Africa	2.49	2.82	3.41	4.13	7.79
Total production	185.45	209.27	238.17	272.16	401.23

This development of the market reflects a high demand for medical devices and implants in clinics and diagnostic centers worldwide. Only in the United States, the number of patients using various IMDs exceeds 25 million [20]. Over two million stents are annually implanted in cardiac surgery and urology. An even higher number of patients use temporary medical devices, such as catheters (400 million) and kidney dialyzers (25 million) [21]. First on the list of the manufactured IMDs are cardiovascular implants (stents, defibrillators, and pacemakers), followed by orthopedic implants. The next group includes neurological stimulators and signal transducers, as well as gastrointestinal bands and laparoscopy systems. Auditory prostheses are the most dynamically developing group of polymer implants with the production prospect of $27.9 billion by the year of 2016. Similar to most high technologies and taking into account the regulations developed by state control institutions, the period from laboratory research to commercial production for the nanomedical innovations is 12–15 years. The first successful results in the field of nanomedicine were obtained in the early 2000s; correspondingly, a considerable increase in the number of nanoscale drugs and devices can be expected as soon as the end of the current decade. By the end of 2010, FDA approved 22 ready-to-use nanopharmaceutical implanted systems, and the number of systems that undergo different phase of clinical trials is even higher. Concurrently, about 25 items of IMDs in Europe are in the final stage of clinical trials.

Note that the commercialization of nanotechnology products is accompanied by exponential growth in the corresponding research and development activities; moreover, as is evident from Fig. 1, the rate of increase in the number of publications (patents and research papers) in the field of nanobiomedicine exceeds that rate of the corresponding publications in the remaining fields of nanotechnology. Note also that the improvement of regulations and legislation for design and use of implants and prostheses, which took place in 1989, had a significant effect on the intensity of the corresponding research, as is reflected by the breakpoint in curve 2, which is denoted for clarity by the corresponding symbol. According to the location within or outside the body, all the currently known implants fall into intracorporeal and intercorporeal. In addition, the devices intended for analysis, diagnosis, and drug delivery are traditionally used under in vivo, ex vivo, and in vitro conditions. Excluding all intercorporeal implants, such as ocular contact lenses, dental prostheses and fillings, periodontal membranes, transdermal systems, wound coverings, and so on from consideration, this review will consider intracorporeal miniature systems except for the traditional orthopedic devices, cartilage implants, surgical meshes, pericardium, and other macroscale medical constructs.

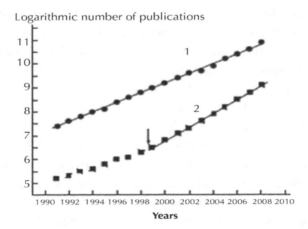

FIGURE 1 Exponential growth in the number of publications (research papers and patents) in (*1*) all nanotechnology areas and (*2*) nanobiomedicine area. Arrow denotes approval datum of the first regulation documents for implants in the United States and 1 year later in Europe.

12.2 MODIFIED CARDIAVASCULAR STENTS AS IMPLANTS PROVIDING DRUG DELIVERY FUNCTION

Cardiovascular diseases are first on the list of mortality an morbidity, exceeding the cancer diseases and casualties. Cardiovascular pathologies are among the basic causes of short lifespan and deteriorating demographic situation. In 2008, the World Health Organization (WHO) acknowledged Romania as the most unfavorable country, with cardiovascular diseases accounting for 52% of fatal cardiac cases. In 2005, Russian Federation headed this list in Europe, with 57% of fatal cases, which is twofold higher than the most trouble-free developed countries [22]. According to the World Health Organization (WHO), 17.3 million people died from cardiovascular related diseases in 2008, and this number is projected to rise to an estimated 23.6 million by 2030. Therefore, it is quite natural that the prevention and therapy of cardiovascular diseases is considered to be a most important priority in the field of demography, health promotion, and increase in the lifespan of population [22, 23].

The latest advance in the field of nanotechnology most pronouncedly appeared in construction of cardiovascular stents. First and foremost, this concerns the main concept in design of the stents that provide a prolonged and local drug delivery. In addition, the obligatory requirement to cardiovascular stents is effective hemocompatibility. The combination of controlled drug delivery and biocompatibility may be attained by nano-structuring the surfaces of cardiac implants, e.g., constructing meso- and nanoporous layers responsible for controllable release, protein adsorption, and adhesion of blood cells (platelets, erythrocytes, leukocytes, etc.).

A typical complication when implanting a vascular stent is restenosis as a result of neointimal hyperplasia, on the one hand, and thrombosis activated by platelets, on the other hand. After the active clinical and technical search for efficient macro and micro constructs (the search for an optimal geometry of bare metal spiral stents) have been depleted, it has become clear that the classic paradigm for introduction of metal devices should be replaced with elaboration of surface modified devices able to dose drugs and provide their prolonged elution. The pioneers in modern technologies for producing next generation cardiac implants with a modified surface (drug eluting stents) are several companies, i.e., Boston Scientific,

Johnson & Johnson, Medtronic, Guidant and the others. Currently, designs of the newest stents with biodegradable or inert coatings provide for the targeted delivery of various cardiac drugs, such as Paclitaxel, Sirolimus, Zotrolimus etc., which are responsible for the inhibition of cell proliferation and, thereby, decrease the incidence rate of restenosis compared to helical metal stents of the first generation.

However, the antiproliferation effect is most pronounced only soon after implanting (the first 6 months); however, as has been recently shown [24, 25], these constructs can initiate thrombosis because of a decrease in cell proliferation and partial degeneration of the neointima [26]. Construction of the nanoporous surface structures with aluminum oxide or titanium oxide for release of Tacrolimus [27], carbon nanoparticles with chromium additions that form the surface layer of the implants for dispensing Paclitaxel [28], and gold [29] is aimed at solving a polyfunctional problem, namely, concurrent provision of (a) biocompatibility of implant and (b) the necessary kinetic profile of drug release. When improving biocompatibility, it is necessary to take into account the topography characteristic of the stent surface (e.g., roughness), which should simulate the natural vascular surface in a nanoscale range. Achieving this effect should lead to improved cell adhesion and stent endothelization and, as a consequence, decrease the probability of thrombosis due to the tight contact between the stent surface and endothelium [30]. The results of constructing the surface structures with nickel titanate [31] or hydroxyapatite [32], which are analogous to the specific features of natural vascular surface, suggest with a certain degree of reserve that one of the promising directions in nanotechnology stent constructs is designing of their nanopatterned surfaces. These surfaces should concurrently provide the following two functions: targeted drug delivery and regulated cell adhesion as the factor that limits restenosis in short terms and the factor that prevents long-term thrombosis as well as promote endotheliazation [33] see the scheme in Fig 2. In the case of specific therapy of vascular tissues and concurrent diagnostics, it is proposed to use magnetized metals as a stent core structure [34]; these metals attract superparamagnetic nanoparticles (e.g., iron oxides) from the bloodstream, thereby allowing not only targeted drug delivery in the absence of external magnetic field, but also to contrast abnormal regions on the vascular surface [35].

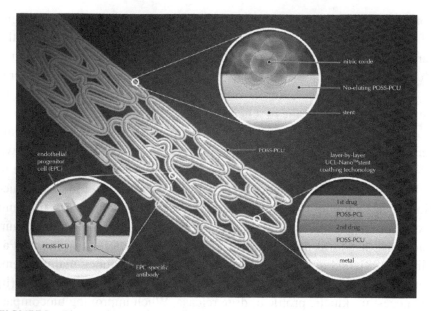

FIGURE 2 The novel generation cardiovascular stents coating with special nanocomposite polymers like POSS-PCU (polyhedral oligomeric silsesquioxane and copolymer carbonate-urea urethane: trade named UCL-NanoTM) has been developed specifically to capture oxide nitrogen and EPC specific antibodies (factor of endothelialization) simultaneously.

Another promising direction in the development of implanted stents is a complete rejecting of metal alloys as basic coiled structures. It is proposed to use for these purposes biodegradable polymeric materials, namely, polylactide-co-glycolides, PLGAs, or poly(3-hydroxybutyrate-co-3-hydroxyvalerate),PHBV. Controlled biodegradation allows, first, for a prolonged drug release to local regions of the cardiovascular system and, second, for a decrease in the risk of repeated thrombosis and hyperplasia with biodegradation of stent [36]. When designing this type of implants, two important conditions should be met, i.e., (1) the rate of stent biodegradation should be approximately equal to the cell proliferation rate and (2) the biodegradation products should not be toxic or induce undesirable immune responses. The mentioned biopolymers almost completely satisfy these criteria, since their intermediate degradation products are members of the physiological Krebs cycle and the final products are molecules of CO_2 and H_2O [37].

12.3 MODERN ORTHOPEDIC IMPLANTS AND THEIR FURTHER DEVELOPMENT

The modern orthopedic implants (OIs) belong to the class of nanocomposites with rather intricate structural organization. They should not only effectively withstand the stresses, but should be also biocompatible and provide for a prolonged targeted drug delivery of biologically active substances. In the general case, OI biocompatibility is determined by the cell response of bone tissues to the implanted material, which characterizes the degree of OI integration into the implanted region. The innovative approach to OI construction as nanobiocomposites is in that the material used to replace bone tissue is represented by a composite with a successive hierarchical structure of spatial elements of organic and inorganic origins. In the nanoscale range, the used structure forming elements are collagen fibers, hydroapatite, and proteoglycans, which together provide for the structural stiffness, stability, and concurrent plasticity of a natural bone.

The effective use of a bone implant is based on its biocompatibility at the cell level. Most researchers who work in this field believe that it is possible to influence the cell behavior via biochemical mediators (with involvement of biologically active substances), surface topography of implant (creation of modifying surface layer or coating), and induction of excitation signals [38]. All of these processes provide a stable cell adhesion and, consequently, form a perfect contact between nonphysiological material and bone tissue. In this situation, the implants maximally biologically mimic the natural bone material. In the absence of such correspondence, the cells in contact with the implant surface change at both the morphological and functional levels, as well as alter their behavior during adhesion, proliferation, endocytosis, and gene regulation. The OI surface is modified using various state-of-the-art technologies that form a nanoscale structure, including nanolithography, phase separation, chemical etching, molecular self_assembly, and electrospinning of micro- and nanofibers.

Nanostructuring of implants creates highly developed surfaces, while the implants acquire fundamentally new mechanical, electric, optical, superparamagnetic, and other characteristics. The set of these properties will allow for the successful application of OI in targeted delivery of low molecular weight drugs, active macromolecules, and cells in the nearest future, as well as their use as diagnostic and biosensor elements. The current concepts of the interactions between the bone tissue and composite materials include adsorption of specific proteins, such as fibronectin and vitronectin, and subsequent adhesion of osteoblasts and other bone cells [39, 40]. In this case, both the composition and structure of the protein layer determine the quality and intensity of cell adhesion and, consequently, the integration ability of bone tissue which determines the differentiation, growth, and death of bone cells on the surface of implant [39, 41].

Currently, intensive studies in this field aim to clarify how the surface energy, chemical composition, charge, morphology, and nanoscale topography of the phase boundary influence the adsorption of specific proteins [42, 43]. In particular, fibronectin predominantly adsorbs on the calcium-phosphate surface of composite [44]. Then, the adhesion intensity of osteoblasts increases in the presence of the amino acid motif of the absorbed protein on the surface, such as arginine-glycine-asparagine (RGD sequence in Fig. 3), as well as when osteoprosthesis is heparinized [45]. In an ideal situation, the protein adsorption layer should stimulate adhesion of the cells that positively influence biocompatibility of the implant, e.g., osteoblasts, and prevent adhesion of fibroblasts, which negatively influence the biocompatibility of implants [39]. As we already mentioned, one of the most important factors in OI biocompatibility, along with adsorption processes, is the surface topography of the implant. This particular factor has a considerable effect on the proliferation and differentiation of osteoblasts, the young cells responsible for the formation of new bone tissue [46]. It is not completely clear how the nanoscale surface topological elements act on the behavior of these cells [47]. However, there are serious reasons to believe that surface structuring in a nanoscale range modulates the cell membrane receptors, thereby modifying the communication between cells [48]. In addition, a nanoscale landscape of the implant surface influences the adsorption and conformational changes of integrin-binding proteins (see Fig. 3), which also activates the cell signaling receptors [49].

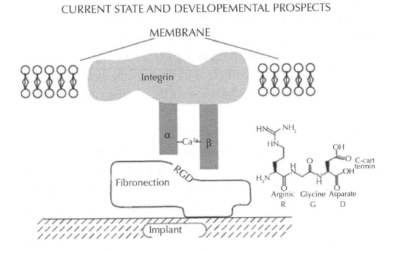

FIGURE 3 Protein adsorption on ostheoimplant surface and following interactions adsorbed proteins with cells of bone tissue.

Nanoscale implant surface topography comprises structural elements, such as pores, corners, slot-shaped hollows, nanofiber structures, nodular intertangling, and their combinations. The specific topographic features of the surface considerably influence its interaction with the cells of surrounding tissue, and this effect on cell adhesion and proliferation is frequently more significant compared to the effect of implant's chemical structure [50]. The situation is somewhat complicated by that individual cell types respond differently to the surface topography. For example, the adhesion of osteoblasts increases with the surface energy of carbon nanofibers, whereas the adhesion of fibroblasts and smooth muscle cells decreases [51]. This is why the detailed mechanism underlying the cell interactions with implant's surface elements is still vague and, in most cases, designers form the topographic landscape in a rather empirical manner.

The surface chemistry of implants plays an important role in their service. During implantation, proteins and other biomolecules adsorb on the implant's surface. The changes in native conformations of biomolecules on a non_physiological surface initiate a trigger cascade of inflammatory (immune) responses, which interfere with the integration of the implant into the living tissue or enhance the formation of fiber capsule. A certain success in the prevention of the negative processes is achieved via a controlled administration of antiinflammatory drugs. The formation of the surface layers, for example, of gold modified with thiol groups and carrying spatially structured functional groups (–CH3, –OH, –COOH, and –NH$_2$) able to bind integrin or regulating cell adhesion, is no less promising [52]. Hydrophilic–hydrophobic balance of the implant surface influences the behavior of an adhered cell.

In addition to modern methods (atomic force microscopy, transmission electron microscopy, etc.), surface wettability (method of contact angle) is used as a simplified rapid assay for hydrophilicity. Amazing as it may seem, a macroscopic characteristic, such as wettability, reflects the OI structure on the nanoscale level. In particular, the surface wetting of aluminum can be increased by decreasing the particle size from 167 to 24 nm [53]. In another work [54], ultrathin titanium crystals displayed a high wettability, which stimulates osteoblasts to proliferation and adhesion and, as a consequence, accelerates the implant integration with the bone tissue. Regarding the bone tissue as a nanocomposite, the current strategy of bone engineering is based on creating a combination of inorganic compounds (calcium triphosphate, hydroxyapatite, organic silicon glass, as well as carbon tubes [55, 56] and organic polymers, such as copolymer of lactic-co-glycolic acid, PLGA; poly(3-hydroxybutyrate), PHB; polycaprolactone; polyphosphazene; chitin; and chitosan. Polymeric matrices form 3D structures for the repair or regeneration of the bone tissue. An original method for OI construction consists of the incorporation of a nanoceramic reinforcing agent that ensures an optimal surface topography and increases the strength of material [55]. The presence of mineral nanoparticles on the OI surface stabilizes its contacts with the cell tissue (integration into tissue) and strengthens the signal transduction of growing cells, thereby enhancing osteogenic activity.

The matrices of the nanofibers impregnated with drugs display the characteristics necessary for growing organs and tissues in cell engineering. Characteristic of the polymer nanofibers produced by electrospinning are high surface to volume ratios, large specific porosity, and adjustable size of nanopores. This set of characteristics makes the nanofiber matrices the most promising material for growing bone and other biological tissues [57, 58]. The ultrathin fiber sizes and their branched surface create optimal conditions for immobilization of growing cells, while the system of pores enhances the metabolic processes associated with cell growth and vital activities [59]. Stem cells are of considerable interest in this connection [60]. Another nanotechnology approach to this problem consists of improving the biocompatibility of the surface. This is attained by coating the implant surface with nanofibers and increasing the flow of drug encapsulated in them [61].

Biocompatibility can be improved by introducing biologically active substances into the modified surface layers of a bone nanoimplant. Controlled release of antiseptics, peptides, extracellular proteins, osteoinductive growth factor, and osteogenic cells is used to stimulate osteogenesis (formation of new bone tissue) and the replacement of biodegradable regions of the implant with bone tissue [62, 63].

Thus, state-of-the-art bone implants should simultaneously contain nanostructured biocompatible surface and the active compounds in their surface layer, which creates an antiseptic effect, as well as controls the state and growth of bone tissue at the boundary with the implant. The adhesion and subsequent colonization of bacteria on the OI surface lead to subsequent infectious complications after implantation. Despite preliminary aseptic procedures, bacterial infections present a serious problem during post surgery period, when IO is in the body. In particular, the rate of complications in hip arthroplasty reaches 3% and the rate of reinfection, even 14%. Antibiotics, the immobilization of antimicrobial agents within the matrix or on its surface, the addition of antiseptic metals (such as silver, copper, or titanium), and the encapsulation of nitrogen oxide are used to prevent the propagation of pathogenic microorganisms [62, 64]. Pioletti et al. [65] have reviewed the latest data on the controlled drug release from OIs. A considerable segment of the innovative research and developments includes nanotechnology modifications of traditional implants, first and foremost, involving titanium and its alloys, in particular with nickel (TiNi). The specific feature of new generation titanium implants is their nanostructured surface. For this purpose, titanium nanotubes vertically oriented relative to the OI surface are currently synthesized to provide the encapsulation of the drug with its subsequent controlled release into the bone tissue. The rate of drug elution from titanium nanotubes can be regulated by the thickness and composition of biodegradable polymer layer used to coat the tube surface and represent an additional barrier for drug diffusion (*see* Fig. 4) [66].

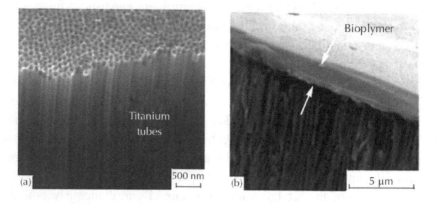

FIGURE 4   Nanostructured surface of a titanium osteoimplant (a) formed as tubes and (b) covered with a layer of poly(lactide-co-glycolide).

Researchers from Brown University (United States) have demonstrated that the quality of bone titanium implants can be considerably improved by coating them with a film of polypyrrole (a conductive polymer). This makes it possible to control and program the elution rate of antibiotics and antiinflammatory drugs by exciting an electric pulse in this electroconductive polymer [67]. Bisphosphonates, correctors of metabolism in the bone and cartilage tissues, prevent the lysis of the bone tissue near the implant and, consequently, enhance its long-term fixation at the implantation site. Several representatives of this class of drugs, e.g., Zoledronate® (both salt and acid variants), pamidronate, and ibandronate, encapsulated in hydroxyapatite nanoparticles display a local prolongation effect for inhibiting the activity of osteoclasts (giant cells) and block calcium desorption from bones as a manifestation of osteoporosis. According to the definition by the FDA [66], in this case, OIs are used as a combined system that comprises a support function and the role of the matrix for controllable drug release.

A factor that complicates the use of nanostructured composite implants and scaffolds is the destruction of their elements in the body. Because of the potential cytotoxicity of the products of implant enzymatic hydrolysis, it is necessary to use polymeric materials that are degradable to nontoxic products, in particular CO_2 and H_2O, as is the situation with the biodegradable polymers of poly(3-hydroxyalkanoates) (PHAs and their derivatives) or poly(α-hydroxyacids) (PGA, PLA, and their copolymers PLGA). The use of carbon in implants carries a certain toxicological risk; however, clearance may be increased via hydrophilization of their walls with the functional groups, such as –COOH, –OH, and –NH_2 [69]. Application of biodegradable materials in the current OIs or scaffolds requires serious studies aimed at regulating the rate of biodegradation [70], since, in an ideal situation, the rate of implant resorption should very precisely coincide with the growth rate of bone tissue in (or its penetration into) OI pores.

In summary, it can be expected that new commercial OIs will appear in clinics in the next years that will differ from the currently available implants in (a) nanopatterned surface and/or matrix; (b) combination of orthopedic and therapeutic functions; and (c) high biocompatibility and antibacterial resistance. Currently, the following well-known manufacturers have brought to the market the composite osteoimplants for replacing bone tissue in knees and hips, in spinal cord surgery, maxillofacial surgery, and so on: Johnson and Johnson, Biomet, Apex Surgical LLC, Encore Orthopedics Inc., Optetrack Inc., Osteoimplant Technology Inc., Smith and Nephew Inc., Stryker Howmedica Osteonics, Zimmer Inc., Sulzer

Orthopedics, and Depuy Co. In 2009, Medtronic Inc. (Minneapolis, United States) announced the production of Mastergraft ceramic scaffolds for growing spinal implants that elute the corresponding drugs. SurModics Inc. (Minnesota, United States) has reached certain success in surface modification and subsequent drug immobilization in the surface layer. However, OIs that meet all three above characteristics are practically still absent and will be the focus of innovative developments in this decade.

12.4 MODERN GENITOURINARY IMPLANTS

Gynitourinary implants (GIs) are widely used for contraception and prolonged therapy (1 year and more) of female pelvic organs, for example, in endometriosis. Specific construction features of such implants are their elasticity, compatibility with adjacent tissues, and, which is most important, drug encapsulatio in their matrix. However, in some cases, specialized laparoscopically introduced microimplants are used to induce a cell rejection response and subsequent growth of fibrous tissue for blocking fallopian tubes, thereby preventing pregnancy.

A polymer-metal device, Essure (Conceptus Inc., San Carlo, California, United States), registered and approved by the FDA, is an example [71]. In 2006, the contraceptive implants Implanon (Organon USA Inc.) and Nexplanon (Merck and Co. Inc., United States) were registered by the FDA and appeared on the market; these microimplants are multi-layered copolymer constructs carrying up to 70 mg of etonogestrel. Unlike the widely known contraceptive rings and their analogs (Norplant and Jadelle), these microimplants provide a daily dose of ~30 μg and more of long-term drug elution (for 3 years) without hurting patient tissues [72, 73]. To visualize their location, these implants are provided with contrasting (X-ray or NMR) characteristics, e.g., by copper doping [74]. Among the latest devices of this type, one should note the Mirena (Bayer, Germany) polysiloxane implant, produced in 2007, which is capable of the controlled release of hormonal compounds for 5 years. The disadvantages of all the currently available commercial GIs are, on one hand, a limited contribution of innovative technologies to their production and, on the other, the lack of the possibility to modify the rate of drug elution according to circadian and monthly body cycles. The current chronobiology recommends synchronizing a targeted drug delivery to the activity of the corresponding organ. A programmed pattern of targeted drug delivery can soon be achieved by implanting microchips into the problem areas of the body. These feedback devices are currently produced by innovative methods (nanolithography, etching, electron micrography, imprinting, etc.). Depending on the changes in biochemical composition of the ambience, these miniature-combined chips provide for regulated drug release, similar to the transdermal and insulin-containing electromechanical systems [75].

Nanotechnology is applied to designing new methods of treating urological diseases in two main directions. When dealing with functional abnormalities, such as bladder dyskinesia, prostatitis, and hyperplasia, various types of nanoparticles are used as diagnostic and therapeutic agents. The second direction of application of nanobiomedicine in urology is the reconstruction of lost or congenitally abnormal elements of the genitourinary system using cell-engineering methods. This involves the restoration of ureters, bladder repair, cosmetic surgery of genitals, and so on. In this review, the latter aspect associated with the implantation of reconstituted urological elements and tissues is of interest. In this case, a key role of cell engineering is evident, namely, the growth of cells, tissues, and elements of organs under special conditions and with the help of specialized 2D and 3D matrix systems (scaffolds). Simplified sketch of tissue growth on matrices is displayed in Fig. 5.

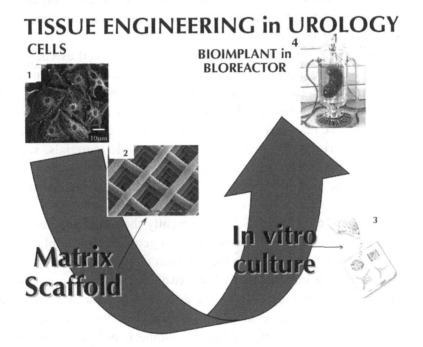

FIGURE 5 The principal stages of tissue engineering for urology (1). Compilation elements adopted from Ref. [76] (2). Perspectives on the role of nanotechnology in bone tissue Engineering [77] (3). Spheroid culture as a tool for creating 3D complex tissues. http://dx.doi.org/10.1016/j.tibtech.2012.12.003. [78] (4). Recent advances in the field of tissue bioengineering have taken scientists closer to growing organs in the laboratory. Cit. David Mack/Science Photo Library.

Cell technology today has started to compensate for the deficiencies in tissues acquired due to traumas or surgery. In particular, Pattison et al.

[79] pioneered the construction of a nanoscale biodegradable 3D matrix for the successful growth of bladder smooth muscle cells. The fibrinogen nanofibers produced by electrospinning were used to design a new type of porous matrices to produce cell material in urology [80]. Bioartificial (hybrid) implants as analogs of the kidney [81] are developed at the University of Michigan (United States). Their working element is a nanoporous silicon membrane that concurrently works as a filter and a matrix for immobilization of renal tubular cells. These devices are intended for the cases of acute kidney failure. As is mentioned above, a drug with antiseptic and cell stimulating functions can be immobilized in a polymeric material, e.g., biodegradable poly(lactide-co-glycolide).

Trauma, toxic or inflammatory circumstances as well as congenital diseases of a genitourinary tract may result in serious damages and even the loss of function. In this situation novel relevant biomaterials are crucially demanded to substitute damaged organs. For example the search for an alternative reservoir to replace the native bladder has proved to be crucially intricate. Synthetic, non-biodegradable materials, such as polyolefines, silicone, rubber, polytetrafluoroethylene and others were initially used in urological surgery. However, these genitourinary implants quickly covered with a crust, were prone to infection and subject to host–foreign body reactions [82]. Currently the tissue engineering is play an increasing role in the management of genitourinary implant variety. Scaffolds for cell cultivation may use not only biopolymers but also artificial tissues or combination of the two. In particular, a combined scaffold of collagen and polyglycolic acid even enhances even tissue regeneration [83]. Sharma et al. [84] performed bladder augmentation in rats with a special polymeric scaffold (elastomeric poly(1,8-octanediol-co-citrate)). The polymer matrix was seeded on opposing surfaces with human mesenchymal stem cells (MSC) and urothelial cells (UC). At 10 weeks morphometry of MSC-polymer-UC sandwiches revealed muscle/collagen ratios approximately 2 times more than controls. Additionally the data of work [83] displayed that MSC colony support partial regeneration of bladder tissue in vivo, A new approach in the area of biomaterials also for the lower urinary tract could be the use of electrospun nanofibers which reveal a superior viability in comparison to other materials, most probably due to their high surface/volume ratio imitating an extracellular matrix structure. Growth of stem cells on the nanofiber material and differentiation into the different cell types has been confirmed in histological analysis [85].

To make the picture complete, the innovative works in the field of urological surgery should be mentioned. In particular, an EnSeal™ implant (SurgRx, Palo Alto, California, United States) used in laparoscopic resection of the prostate and comprising millions of nanoparticles, which enhances healing injured small veins, thereby minimizing the tissue damage in the involved organ [86]. Submicron tweezers, a potential urological surgery tool, is now used in surgery of spermatic ducts or, e.g., in vasectomies [87]. High-molecular-weight implants, such as hydrogels or elastic polymeric constructs, are used when treating and replacing genitourinary elements. Hydrogel membranes are formed for a number of implanted devices to separate the ambience and drug container. The therapy of prostate tumors, as well as the regulation of adolescent sexual maturation, involve the administration of hormones; for this purpose, the Hydron, Supprelin, and Vantas gel implants with about 1-year-long drug elution have been developed [88]. The reservoir systems that utilize an osmotic pump principle, i.e., ALZET (Alza) [89] and DUROS [90], have been designed; their separation membranes are formed of cellulose ester and polyurethane, respectively, and they elute at a constant rate a protein, LHRH agonist, administered to treat prostate cancer.

Biodegradable polymers represent a separate group of polymer implants in urology. By analogy with bioresorbable suture materials and nanoparticles, the following groups of natural polymers have been used for their production: poly(α-hydroxyacids) (polylactides and their copolymers) and poly(β-hydroxyalkanoates) (PHAs) and the copolymers with hydroxyvalerate, hydrooxyhexanoate, etc. An example is the multifunctional implanted system Zoladex™ (Astra Zeneca, Canada), which was designed over 20 years ago and is currently widely used in clinical practice for the long-term therapy of endometriosis, breast tumors, and prostate tumors [91]. Note that polyhydroxyacids (PLA and PGLA) and poly(ortho esters) were certified by the FDA US in 2009. Catheters of various types and diameters are now widely used as transient implants. They are temporarily introduced into various body parts, such as the bladder, to replace ureters, to enhance the therapy of renal tissues, and to prevent gall bladder dyskinesia. The main requirements to polymeric catheters are nontoxicity; biocompatibility; and, in particular, their antiseptic effect. For this purpose, similar to vascular catheters, their surfaces are modified and enriched for encapsulated drugs, most frequently, a bactericidal agent. The controlled delivery of the drug desorbed from the walls of an implant in this situation allows for the avoidance of numerous complications caused by urine-containing, bacterially active media.

12.5 NANOIMPLANTS WITH ENCAPSULATED DRUGS FOR OPHTHALMOLOGY

The eye diseases present a serious problem, first and foremost, for the population of industrially developed countries. These diseases include age related retinal changes, diabetic retinopathy, structural alterations in the lens, and so on. A systemic drug delivery to the anatomically intricate organ, such as the eye, and especially to the intraocular elements is not a simple task, which interferes with the therapy of the uvea, vitreous body, and optic nerve. Because of multiple tissue barriers and lacrimal drainage, only 1–3% of the active substance of traditionally administered drugs (e.g., as eye drops) reach the problem areas within the eye [83, 92]. Therefore, it is planned in the nearest future to widely use nanoparticles and polymer implants with encapsulated drugs for prolonged and targeted drug delivery. The eye lens spatially separates the eyeball into the anterior segment, which comprises a cornea, sclera, and conjunctiva, and the posterior segment, with all of its elements localized beyond the lens deep in the eye (retina, vascular membrane, vitreous body, and optic nerve). This is why the ocular implants are classified according to the eye anatomical structure into anterior ocular and (e.g., contact lenses) and intraocular devices.

The installation of ocular micro- and macroscale implants is associated with a number of issues that should be briefly mentioned in this section. First, the eye is a paired organ; correspondingly, this provides for an objective comparison of the outcome of innovative treatment, be it a surgery or therapy, of one eye with the intact state of the other eye, using it as a control [94]. Second, the eyeball structure is rather transparent, which allows for an efficient use of optical and photosensitive methods in diagnosis and control. Third, the anatomical structure of the eye includes miniature elements in both the anterior (in front of the lens) and posterior (beyond the lens) chambers. Correspondingly, the ocular implants are limited by small sizes, including their nanoscale geometry. Fourth, the loss of sight is not necessarily associated with the loss of life, but considerably complicates it; therefore, the efforts for expanding the market of ocular preparations are very active, and the ophthalmological market by 2011 had reached several billions of dollars.

The current state and nearest future of implants in ophthalmology are to a certain degree reflected by Tables 2 and 3, which lists the IMDs for local drug delivery both commercially produced and at various stages of clinical trials. All the tools are partitioned according to the target of their prolonged delivery, namely, the anterior eye chamber (cornea, conjunctiva, sclera, etc.) and the posterior chamber (retina, optic nerve, vitreous body, macula, etc.). According to the current trends, the analogous drug forms already tested in ophthalmology but designed for gene therapy will be further developed. In particular, it has been shown at the initial stage of clinical trials in the United States that some RNAs (such as low molecular weight interfering RNAs) implanted into biodegradable polymer complexes display a high efficiency in treatment of the congenital retinal degeneration and macular degeneration [95]. In addition, it is purposeful to focus on the design and production of the drainage systems for treatment of glaucoma [95], one of the most widespread diseases. In an ideal situation, these systems (ANDIs) should contain the polymer lines for drainage of lacrimal fluid and exudate and provide for a targeted delivery of the drugs decreasing intraocular pressure and preventing destruction of the optic nerve. Advance in designing the new generation implants for the anterior eye chamber will allow the miniature sensors for biochemical eye surface monitoring [96] and concurrent drug elution [97] to be installed directly into contact lenses or into posterior compartment (*see* Fig. 6).

TABLE 2 Promising implanted devices for local drug delivery to the anterior eye chamber [65].

Active agent	Trademark	Type	Matrix	For treating of	Stage of development
Betaxolol	Betoptic®	Eye drops	Amberlite IRT-69	Glaucoma	In market
Azithromycin	AzaSite®	Eye drops	Polycarbophil	Bacterial conjunctivitis	In market
Ketotifen		Soft contact lenses	–	Allergic conjunctivities	Phase 3
Latanoprost		Tampon cylinder	PVA and cellulose	Glaucoma	Phase 2
Latanoprost		Implant of eye connective tissue (subconjuctival)	PVA-PLGA composite	Glaucoma	Phase 2
				Keratoconjunctivitis	Phase 3
Cyclosporine (LKh-201)		Implant in the eye episclera	Silicon	Uveities	Phase 2
Dexamethasone phosphate EGP-437	EyeGate 11®	Device for microionophoresis	–	Difficulties in lacrimation (dry eye symptom)	Phase 3

TABLE 3 Promising implanted devices for local drug delivery to the posterior eye chamber [65].

Active agent	Trademark	Type	Matrix	For treating of	Stage of development
Ganciclovir	Vitrasert®	Intravitreal implant	Poly(ethylene-co-vinyl alcohol)	Viral retinities (inflammation of retina)	In market
Fluocinolone acetonide	Retisert®	Intravitreal implant	PVA-silicone	Uveitis	In market

TABLE 3   *(Continued)*

Fluo-cinolone acetonide	Iluvien®	Intravitreal implant	Polyimide-PVA	Diabetic macular edema	Phase 3
Dexametha-sone	Ozurdex®	Intravitreal implant	Biodegradable PLGA	Renal vonous congestion	In produc-tion
Brimonidine	–	Intravitreal implant	Biodegradable PLGA	Age-related macular de-generation	Phase 2
Triam-cinolone acetonide	I-vation™ TA	Intravitreal implant	PMMA/eth-ylvinyl acetate	Diabetic macular edema	Phase 2
Triam-cinolone acetonide	IBI-20089	Intravitreal implant	Oil medium	Renal venous congestion	Phase 1
Triam-cinolone acetonide	RETAAC	Intravitreal implant	Biodegradable PLGA	Diabetic macular edema	Phase1

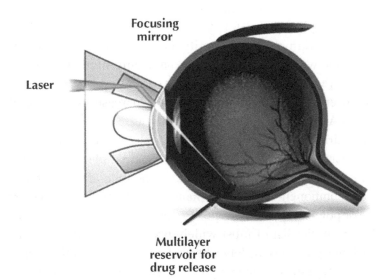

FIGURE 6   Photodynamically activated system for controlled drug delivery within the intraocular region of eye.

The current nanomedicine provides the possibilities for replacement of several eye elements with artificial implants. The replacement therapy in ophthalmology includes the repair of the retinal nerve (formation of polymer conduit) and optical nerve [95]. A separate field in cell engineering utilizes 2D and 3D biodegradable scaffolds that enhance cell proliferation and function for the growth of eye tissues and, in the nearest future, also elements of the eye, e.g., the cornea [98]. In this case, the surface topography of the polymer should meet special requirements, since the nanostructured surface of the scaffold provides controllable cell adhesion and migration. Thus, we again meet the problem of controllable drug elusion from polymer matrix. For example, the delivery of the peptides capable of spatial self-organization and self-assembly will be used in the nearest future for the regeneration of the retinal axon [99].

12.6 PROSPECTS OF NEUROLOGICAL IMPLANTS

Neurology is an interesting and most promising field for implant application. Typical implants used for this purpose are polymer conduits intended for the restoration of the integrity and conductivity of nervous conductors. It is known that the nerve cells have only limited ability to repair and the nerve tissue has a limited ability to regenerate after injuries, e.g., craniocerebral traumas. Moreover, an increase in the rate of diseases, such as Alzheimer's or Parkinson's disease, as well as of cerebrovascular impairments and subsequent dystrophy of nerve fibers, requires an ever-increasing number of micro- and nanoimplants providing for transduction of the signals in the central and peripheral nervous systems. The major component for these implants is biodegradable polymers of polylactide or polyoxyalkanoate type. The prospects for nanoimplants in this field, third in the list of most widely used devices (after cardiac implants and bone prostheses), have been detailed in several reviews [100, 101]. Here, organic silicon (polymer) systems are the most promising. Useful properties of structured silicon dioxide have been recently discovered; its nanoporous surface provides a more stable contact with neurons compared to smooth surfaces and prevents tissue overproliferation (gliosis) [102]. The pioneer experiments have demonstrated that the activation of semiconductor nanoparticles with light pulses stimulates rat brain neurons, which allows for the contactless (electrode-free) excitation of a neuronal signal [103]. In this field of regeneration medicine, the implanted materials should be polyfunctional, namely, should work as flexible nanoscale backbone constructs, enhance excitation of signal transduction, and concurrently elute a drug that activate development of a nerve impulse.

12.7 MINIATURE ELECTROMECHANICAL IMPLANTS FOR PROGRAMMED DELIVERY OF BIOLOGICALLY ACTIVE SUBSTANCES

Innovative developments in the chemistry of polymers and supramolecular chemistry, composite engineering, and design of micro- and nanoscale electromechanical systems (MEMS and NEMS) are directed towards obtaining of fundamentally new therapeutic systems for controllable release of low-molecular-weight drugs, macromolecules, and cells [104, 105]. Polymer and inorganic implants equipped with electronic chips create kinetic profiles with different complexities to release the biologically active component. Several implanted systems additionally contain mechanical manipulators, drives, and nanocantilevers. The combination of computer, electronic, and mechanical elements in an implant determines that it is a principally new electromechanical device (MEMS/NEMS) [106–108]. The miniaturization of these devices, which utilize state-of-the-art technologies (nanolithography and imprinting), allows these implants to be considered an independent universal class of the systems for targeted drug delivery [109–111]. The introduction of MEMS/NEMS as drug-delivery-programmed systems into clinical practice is determined to a considerable degree by advances in a new field of biomedicine, personalized medicine, which is a customized therapy of each patient that takes into account his/her individual genetic features and specificity of biochemical processes [112, 113]. The information about symptomatology of disease and data on the metabolites assayed with biomarkers are input into a chip to provide a programmed release of a set of drugs from a system of containers according to these integrate data. Here, the most impressive results are attained when using an integrated treatment of complex pathologies, such as tumor diseases, liver cirrhosis, and blood disorders.

Using a feedback principle, an MEMS is used to design a vitally important organ, the artificial pancreas, which monitors the sugar level in the blood and releases the necessary dose of insulin according to the sugar concentration [114–116]. Currently, there are three hindrances to the commercialization of this system for diabetes control, namely, (a) a reliable programmed device; (b) glucose sensors stable in their characteristics and able to stably function for a long time in the specific aggressive medium, such as blood; and (c) still high price of the final product [117]. Nonetheless, taking into account the high rate of diabetic diseases, in the next 10–12 years, these implanted systems should become a mass therapeutic tool.

Once the first two problems are associated with advances in nanotechnology, the third issue has a rather moral and ethical aspect. No matter how large the expenditures for treating a patient using precision and expensive electromechanical platforms, an improvement in his/her quality of life cannot be limited by material and financial parameters. The last statement may be illustrated by one rare disease with a very limited number of cases. It is known that metabolic disorders require constant therapy to sustain a patient's comfort. In this situation, implanted NEMS devices for the controllable release of deficient enzymes find their own niche in the world pharmaceutical market. For example, Genzyme Therapeutics (United States) [117] specializes in the production of a Cerezyme implant, which elutes recombinant proteins for the long-term therapy of the Gaucher disease. Despite that this disease is very rare (1 per 200,000 in the population), in 2008, this company acquired $1.2 billion from selling this device, which accounts for 27% of its annual profit. As in the case with the NEMS implants, large expenditures for implantation and therapy ($200,000 USD annually per patient) [112] are completely justified, since any alternative method that alleviates suffering and prevents a preliminary fatal outcome is absent.

The specific construction features of the MEMS-and NEMS-type implants include not only their miniature size and stable operation, but also the use of bioinert and biocompatible materials. In their operation principle, these devices are miniature reservoirs that contain a drug, which is eluted in response to electrical signal or movement of piston/micropump walls. The common characteristics for the main types of electromechanical implants are listed in Table 4.

TABLE 4 Comparative characteristics of nanoimplants utilizing different drug release mechanisms.

Principle of drug release	Main characteristics	Therapeutic potential
Implanted electronic chip	Limitations on volume of device only for highly active compounds high probability of tissue capsule formation stricter requirements to bioinert and nontoxic materials in long-term operation invasive implantation procedure high price	Local and more economical drug delivery precise and time-controllable dgur delivery more preferable than parenteral drug administration possibility of pulse and more complex kinetic profile of drug release
		Has advantages as compared eith aerosols and oral administration Combined therapy and diagnosing (monitoring)
Implanted micropump	Limitation on volume of divice only for highly active compounds exclusively for release in liquid from (difficulties in releasing polypeptides due to their low stability in aqueous medium)	Local and more economical drug delivery more preferable than parenteral drug administration possibility of pulse and more complex kinetic profile of drug release;
	Possible structural changes in active component because of shear stress	Has advantages as compared with aerosol, oral, and parenteral drug administration routes
	Precipitation and subsequent plugging of pump piston with the drug presipitated from solution Invasive implantation procedure considerable expenditures for implantation and production	

TABLE 4   *(Continued)*

Implanted polymer chip	Low invasive	Local delivery provides advantages compared to traditional drug injection and aerosol administrations perfect alternative to parenteral drug administration tolerance to delivery via systemic or local administration routes
	Limitations on volume of device only for highly active compounds stricter requirements to drug stability (absence of hydrolysis)	
	Broad therapeutic window difficulties in precise dosing	
		Biodegradable elements do not require repeated invasion for extracting worked out system
Implanted selective membranes	Limitations on volume of device only for highly active compounds requires surgical intervention risk for stoppage of membrane pores with loss of production capacity and selectivity appropriate only for dgur solutions favorable conditions for cell material	Due to local delivery, has advantage as compared with traditional drug injection and aerosol administration
		Perfect alternative to parenteral drug administration depending on implantation site, allow for controlled delivery via bloodstream or to local body area does not require repeated invasion for extracting worked out system

At the micro- and nanoscales, several electromechanical devices for controllable drug release act as pumps with either a drive or by the effect of osmotic forces [118]. In the latter case, there is no need for any electromechanical drive, and strictly speaking, such devices cannot be regarded as MEMS/NEMS. Only a very limited number of implanted systems that utilize a micropump principle is manufactured in batches. The examples include a MiniMed Medtronic Pump for controlled insulin release and SynchroMed II Programmable Pump (commercially available since 1988) for critical therapy (a high level of pain and neurological indications). A series of the Debiotech pumps, which utilize the piezoelectric effect in their design, allows the drug solution to be eluted at a rate of 100 μl/h [119]. Another operating principle of a micropump is to use materials with a shape memory effect [119]; they allow the therapeutic rate of drug delivery to be reduced to 5 μl/h.

12.8 CONCLUSION

The transition from a macroscopic description of implanted systems to molecular and submolecular consideration is a necessary condition for the development of new generation bioimplants. Nowadays, there are good grounds to believe that the structuring of the implant surface in a nanoscale range modulates cell membrane receptors that modify the communication between cells and has a positive effect on adsorption and conformational changes in the proteins responsible for immune and toxic responses. The successive consideration of the cardiac implants based on the example of new generation stents and bone implants, as well as ophthalmological, urological, and others, demonstrates that the combination of the following three factors determines the nearest prospects in implantology: (1) topography and chemical composition of the surface corresponding to the natural organ; (2) the ability to perform mechanical, optical, ion conducting, and other functions characteristic of the replaced element; and (3) the ability to encapsulate drugs with the possibility of their controllable (programmed) and prolonged targeted delivery to a local area of the body.

ACKNOWLEDGMENT

The work was supported by the Russian Foundation for Basic Research (grant no. 13–03–00405-a) and the Russian Academy of Sciences under the program "Construction of New Generation Macromolecular Structures" (03/OC-13) [68, 76–78, 93].

KEYWORDS

- **Miniature electromechanical systems**
- **Nanoscale biochips and sensors**
- **Poly(-hydroxy acids)**
- **Poly(-hydroxyalkanoates)**

REFERENCES

1. Logothetidis, S. *NanoScience and Technology.* **2012**, *61*, 1–26.
2. Huebsch, N.; Mooney, D.J. *Nature.* **2009**, *462, 7272*, 426–432.
3. Hasirci, V.; Vrana, E. Zorlutuna, P.; Ndreu, A. Yilgor, P. Basmanav, F.B.; Aydin, E. *Journal of Biomaterials Science, Polymer Ed.* **2006**, *17(11)*, 1241–1268.
4. Bazhenov, S. L.; Berlin, A. A.; Kul'kov, A. A.; and Oshmyan, V. G.; *Polimernye kompozitsionnye materialy.Prochnost'i tekhnologiya* (Polymer Composite Materials. Durability and Technology), Rus. Edition. Dolgoprudnyi: Intellekt, **2010**.
5. Jorfi, M.; Roberts, M. N.; Foster, E. J.; Weder, C. *ACS Applied Materials and Interfaces.* **2013**, *5(4)*, 1517–1526.
6. Frima, H. J.; Gabellieri, C.; Nilsson, M. -I. *Journal of Controlled Release.* **2012**, *161(2)*, 409–415.
7. Drug Delivery Systems US Industry Study with Forecasts for 2012 and 2017. Study #2294 | March **2008**, 338. *The Freedonia Group.* Cleveland, OH. USA. www.freedoniagroup.com
8. Staples, M. *Wiley Interdisciplinary Reviews: Nanomedicine and Nanobiotechnology* **2010**, *2(4)*, 400–417.
9. Leucuta, S. E. *Current Clinical Pharmacology* **2012**, *7(4)*, 282–317.
10. Stevenson, C. L.; Santini Jr. J. T.; Langer, R. Advanced Drug Delivery Reviews. **2012**. in press doi: 10.1016/j.addr.2012.02.005. Reservoir-Based Drug Delivery Systems Utilizing Microtechnology.
11. Sirivisoot, S.; Pareta, R. A.; Webster, T. J. *Recent Patents on Biomedical Engineering* **2012**, *(1)*, 63–73.
12. Global Industry Analysts Inc.; Nanomedicine: *A global strategic business report,* **2009**.
13. Murday J. S. et al. Nanomedicine: *Nanotechnology, Biology, and Medicine,* **2009**, *5*, 251–273.
14. Ekdahl, K. N.; Lambris, J. D.; Elwing H.; Ricklin D.; Nilsson P. H.; Teramura Y.; Nicholls Bo Nilsson I. A. *Advanced Drug Delivery Reviews* **2011**, *63*, 1042–1050.
15. Bonartsev A. P.; Boskhomedgiev A. P.; Iordanskii A. L. et al. In: Kinetics, Catalysis and Mechanism of Chemical Reactions. From pure to applied science. V.2 -Tomorrow and perspectives. (eds. R. M. Islamova; S. V. Kolesov.; G. E. Zaikov) Nova Science Publishers. New York **2012**. Ch.*27*, 335–350. *Degradation of poly(3-hydroxybytyrare) and its derivatives: Characterization and kinetic behavior.*
16. Iordanskii A. L.; Rudakova T. E.; Zaikov G. E. *Interaction of Polymers with Bioactive and Corrosive Media. Ser. New Concepts in Polymer Science.* VSP Science Press. Utrecht –Tokyo Japan. **1994**, 298.
17. Directive 2007–47–EC of the European Parliament and of the Council.pdf http://www.fda.gov/MedicalDevices/DeviceRegulationandGuidance/Overview/ClassifyYourDevice/ucm051512.htmhttp://laws-lois.justice.gc.ca/eng/regulations/C.R.C.;_c._870/index.html
18. Market Report: World Medical Devices Market September 2007. Acmite Market Intellegence. http://www.acmite.com/market-reports/medicals/world-medical-devices-market.html.

19. Report of Ministry of Healthcare and social development RF Review [online]. http://www.mzsrrf.ru/press_smi/388.html.
20. The World Health Organization Report, **2006** [online]. – http://www.who.int/entity/whr/2006/whr06_en.pdf. b) The World Health Organization Report, 2011. Cardiovascular disease. http://www.who.int/cardiovascular diseases/en/ document: Conception of demography development in Russia until 2015. http://www.demoscope.ru/weekly/knigi/koncepciya/koncepciya.html
21. Kastrati, A. et al. N. Engl. J. Med. **2007**, *356,* 1030.
22. Stone, G. W. et al. N. Engl. J. Med. **2007**, *356,* 998.
23. Lagerqvist, B. et al. N. Engl. J. Med. **2007**, *356,* 1009.
24. Wieneke, H. et al. Catheter Cardiovasc. Interv. **2003**, *60,* 399.
25. Bhargava, B. et al. Catheter Cardiovasc. Interv. **2006**, *67,* 698.
26. Erlebacher, J. et al. Nature **2001**, *410,* 450.
27. Caves, J. M.; Chaikof, E. L. J.; Vasc. Surg. **2006**, *44,* 1363.
28. Samaroo, H.D. et al. Int. J. Nanomedicineю **2008**, *3,* 75.
29. Tana, Y. Farhatnia; Achala de Mel; J. Rajadas; M. S. Alavijeh; A. M. Seifalian. *Journal of Biotechnology* **2013**, *164,* 151–170. Inception to actualization: Next generation coronary stent coatings incorporating nanotechnology.
30. Godin, B.; Sakamoto, J. H.; Serda, R. E.; Grattoni, A.; Bouamrani, A.; Ferrari, M. *Trends in Pharmacological Sciences*. **2010**, *31(5),* 199.
31. Rosengart, A. J.; Kaminski, M. D.; Chen, H.; Caviness, P. L.; Ebner, A. D.; Ritter, J.A. J. Magn. Magn. Mater. **2005**, *293,* 633.
32. Cregg, P.J.; Murphy, K.; Mardinoglu, A.; Prina-Mello, A. Journal of Magnetism and Magnetic Materials. **2010**, *322,* 2087.
33. Zilberman, M.; Nelson, K. D.; Eberhart, R. C. Mechanical Properties and In Vitro Degradation of Bioresorbable Fibers and Expandable Fiber-Based Stents. Published online 30 June **2005** in Wiley InterScience (www.interscience.wiley.com). DOI: 10.1002/jbm.b.30319
34. Yun-Xuan Weng; Xiu-Li Wang; Yu-Zhong Wang. *Polymer Testing*. **2011**, *30,* 372.
35. Kumbar, S.G Kofron, M. D.; Nair, L. S.; Laurencin, C. T. Cell behavior toward nanostructured surfaces. In: Gon-salves KE, Laurencin CT, Halberstadt C, Nair LS: eds. Biomedical Nanostructures. New York: *John Wiley & Sons*; **2008**, 261–295.
36. Balasundarama, G.; Webster, T. J.; J. Mater. *Chem.* **2006**, *16,* 3737.
37. Schakenraad, J. M. Biomaterial Science. Eds. Ratner, B. D.; Hoffman, A. S.; Schoen, F. J.; Lemons, J. E. San Diego: Academic Press, Inc.; **1996**, 140.
38. Kennedy, S. B.; Washburn, N. R.; Simon, C. G.; Amis, E. J. Biomaterials. **2006**, *20,* 3817.
39. Hersel, U.; Dahmen, C.; Kessler, H. Biomaterials. **2003**, *24,* 4385.
40. Feng, Y.; Mrksich, M. Biochemistry, **2004**, *43,* 15811.
41. El-Ghannam, A.; Ducheyne, P.; Shapiro, I. M. J. Orthop. Res. **1999**, *17,* 340.
42. Sagnella, S.; Anderson, E.; Sanabria, N.; Marchant, R. E.; Kottke-Marchant, K. Tissue *Eng.* **2005**, *11,* 226.
43. Anselme, K.; Bigerell, M. J. Mater. Sci.: Mater. *Med.* **2006**, *17,* 471.
44. *Curtis, A.* IEEE Trans. Nanobiosci. **2004**, *3,* 293.
45. Price, R. L.; Ellison, K.; Haberstroh, K. M.; Webster, T.J.; *J. Biomed. Mater. Res.;* **2004**, *70,* 129.

46. Webster, T. J.; Schadler, L. S.; Siegel, R. W.; Bizios, R. Tissue Eng. **2001**, *7*, 291.
47. Curtis, A.; Britland, S. Surface modification of biomaterials by topographic and chemical patterning. In: *Advanced Biomaterials in Biomedical Engineering and Drug Delivery Systems.* Ogata, N., Kim, S.W., Feijen, J., Okano, T., eds. Tokyo: Springer-Verlag; **1996**, 158–167.
48. *Woo, K. M.; Chen, V. J.; Ma P. X.* J Biomed Mater Res. **2003**, *67A (2)*. 531.
49. *Keselowsky, B. G.; Collard, D. M.; Garcia, A. J.* J Biomed. Mater. Res. **2003**, *66A (2)*. 247.
50. Webster, T. J.; Ergun, C.; Doremus, R. H.; Siegel, R. W.; Bizios, R.; *J. Biomed Mater Res.* **2000**, *51, 3,* 475.
51. Faghihi, S.; Azari, F.; Zhilyaev, A. P.; et al. *Biomaterials* **2007**, *28 (27)* 3887.
52. Khan, Y.; El-Amin, S. F.; Laurencin, C. T.; Conference Proceedings of the IEEE Engineering in Medicine and Biology Society.; *1.*; New York, **2006**, 529.
53. Harrison, B. S.; Atala, A. Biomaterials **2007**, *28 (2)*. 344.
54. Li W. J.; Laurencin, C. T.; Caterson, E. J.; Tuan, R. S.; Ko F. K. *J. Biomed. Mater. Res.* **2002**, *60,* 613.
55. Nukavarapu, S. P.; Kumbar, S. G.; Nair, L. S.; Laurencin, C. T. Biomedical Nanostructures; Eds. Gonsalves, K. E.; Laurencin, C. T.; Halberstadt, C.; Nair, L. S. (NewYork: Wiley, **2008**) 377.
56. Kumbar, S. G.; Nukavarapu, S. P.; Roshan, R.; Nair, L. S.; Laurencin, C. T.; *Biomed Mater* **2008**, *3*, 1.
57. Pelled, G.; Tai, K.; Sheyn, D.; Zilberman, Y.; Kumbar, S. G.; et al. *J. Biomech* **2007**, *40*, 399.
58. Kumbar, S. G.; Nair, L. S.; Bhattacharyya, S.; Laurencin, C. T.; *J Nanosci. Nanotechnol.* **2006**, *6*, 2591–2607.
59. Nablo, B. J.; Rothrock, A. R.; Schoenfisch, M. H.; *Biomaterials* **2005**, *26*, 917.
60. Simchi, A.; Tamji, E.; Pishbin, F.; Boccaccini, A. R.; *Nanomedicine. Nanotechnology, Biology, and Medicine.* **2011**, *7*, 22.
61. Popat, K. C.; Eltgroth, M.; LaTempa, T. J.; Grimes, C. A.; Desai, T. A.; *Small* **2007**, *3*, 1878.
62. Pioletti, D. P.; Gauthier, O.; Stadelmann, V. A.; et al. Orthopedic implant used as drug delivery system: Clinical situation and state of the research. *Current Drug Delivery*, **2008**, *5*, 59–63.
63. Gulati, K.; Ramakrishnan, S.; Aw, M. S.; Atkins, G. J.; Findlay, D. M.; Losic, D. Acta Biomaterialia. **2012**, *8,* 449.
64. Sirivisoot, S.; Pareta, R.; Webster T. J. Electrically controlled drug release from nanostructured polypyrrole coated on titanium. *Nanotechnology* **2011**, *22*, 085101 doi:10.1088/0957–4484/22/8/085101. http://www.fda.gov/oc/combination/.
65. Laurencin, C. T.; Kumbar, S. G.; Nukavarapu, S. P.; *Interdiscipl. Rev. Nanomed. Nanobiotechnol.* **2009**, *1*, 6.
66. Ivantsova, E. L.; Kosenko, R. Yu.; Iordanskii, A. L.; et al. *Polymer Science, Ser. A,* **2012**, *54(2), 87–93*
67. Shavell, V. I.; Abdallah, M. E.; Shade Jr G. H.; Diamond, M. P.; Berman, J. M.; *J Min. Invasive Gynecol.* **2000**, *16 (1),* 22.

68. Shulman, L. P.; Gabriel, H. Contraception **2006**, *73*, 325.
    http://www.merck.com/newsroom/news-release-archive/prescription-medicine-news/2011_1109.html.
69. Correia, L.; Ramos, A. B.; Machado, A. I.; Rosa, D.; Marques, C. *Contraception* **2011**, in press , doi:10.1016/j.contraception.2011.10.011.
70. Staples, M. Microchips and controlled release drug reservoirs. *WIREs Nanomed Nanobiotechnol* **2010**, *2*, 400–417.
71. Saiz E.; Zimmermann E. A.; Lee J. S.; Wegst U. G. K.; Tomsi, A. P. Dental Materials **2013**, *29*, 103–115.
72. Fennema, E.; Rivron, N.; Rouwkema, J.; van Blitterswijk, C.; and de Boer, J.; *Trends in Biotechnology*, **2013**, *31(2)*, 271–282. http://dx.doi.org/10.1016/j.tibtech.2012.12.003.
73. Chopa, N.; Kayes, O.; *Trends in Urology and men's health.* **2013**, *1(2)*, 14–16. Recent advances in the field of tissue bioengineering have taken scientists closer to growing organs in the laboratory. Cit. David Mack/Science Photo Library.
74. Pattison, M. A.; Wurster, S.; Webster, T. J.; Haberstroh, K. M. Biomaterials **2005**, *26*, 2491.
75. McManus, M.; Boland, E.; Bowlin, G.; Simpson, D.; Espy, P.; Koo, H. *American Urological Association Annual Meeting*, May 8–13, **2004**, San Francisco, CA, USA.
76. *Humes, H. D.* Semin. Nephrol **2000**, *20*, 71.
77. Lee, D.; Lee, J. T.; Sheperd, D.; Abrahams, H.; Lee, D.; Preliminary use of the Enseal™ system for sealing of the dorsal venous complex during robotic assisted laparoscopic prostatectomy. *American Urological Association Annual Meeting*, May 21–26, **2005**, San Antonio, TX, USA.
78. Eberli, D.; Yoo, J. J.; Atala, A.; Urologe. **2007**, *A 46*, 32.
79. Sharma, A. K.; Hota, P. V.; Matoka, D. J. et al. *Biomaterials.* **2010**, *31(24)*, 6207–62178.
80. Feil, G.; Daum, L.; Amend, B.; et al. Advanced Drug Delivery Reviews. **2011**, *63*, 375–378. From tissue engineering to regenerative medicine in urology — *The potential and the pitfalls.*
81. Murphy, D. G.; Costello, A. J. Nanotechnology. In: *New Technology in Urology.* (P. Dasgupta et al. eds.) Springer Verlag: London *201* , 268.
82. Cruz, C. G. M. PHD Thesis's. *pH-Triggered Dynamic Molecular Tweezers for Drug Delivery Applications.* Queen's University. Kingston, Ontario, Canada September **2011**, 180.
83. Wright, J. C.; Hoffman A. S. Historical Overview of Long Acting Injections and Implants in *Long Acting Injections and Implants* (Wright J.C.; Burgess D.J. (eds.)), Advances in Delivery Science and Technology, Chapter 2. DOI 10.1007/978-1-4614-0554-2-2,Controlled Release Society, Springer. **2012**.
84. Theeuwes, F.; Yum S.I. Ann Biomed. Eng. **1976**, 343–353. Principles of the design and operation of generic osmotic pumps for the delivery of semisolid or liquid drug formulations.
85. Wright, J. C.; Leonard, S. T.; Stevenson, C. L.; Beck, J. C.; Chen, G.; Jao, R. M.; Johnson, P. A.; Leonard, J.; Skowronski, R. An in vivo/in vitro comparison with a leuprolide osmotic implant for the treatment of prostate cancer. *J Control Release.* **2001**, *75*, 1–10.

86. Duncan, R. *Nature Review.* **2006**, *6*, 688–701 Polymer conjugates as anticancer nano-medicines.
87. Maurice, D. M.; and Mishima, S.; Ocular pharmacokinetics, in *Handbook of Experimental Pharmacology*, Sears, M.L.; Ed.; Berlin_Heidelberg: Springer, **1984**, 16.
88. Kuno, N.; Fujii, S.; *Polymers,* **2011**, *3*, 193.
89. Thomson, H.; Lotery, A.; *Nanomedicine,* **2009**, *4*, 599.
90. Pan, T.; Brown, J. D.; Ziaie, B.; *IEEE Eng. Med. Biol. Soc.;* **2006**, *1*, 3174.
91. Parviz, B. A.; Shen, T. T.; Ho, H.; *Invest. Ophtalmol. Vis. Sci.;* **2008**, *49*, Abstr. 4783. Functional contact lens with integrated inorganic microstructures. www.arvo.org
92. Kapoor, Y.; Thomas, J.C.; Tan, G.; John, V.T.; Chauhan A, *Biomaterials,* **2009**, *30*, 867.
93. Liu, B. S.; Yao, C. H.; Hsu, S. H.; et al.; *J. Biomater. Appl.;* **2004**, *19*, 21.
94. Elder, J. B.; Liu, C. Y.; Apuzzo, M. L. J. Neurosurgery in the realm of 10–9, Part 2: *Applications of nanotechnology to neurosurgery—present and future. Neurosurgery.* **2008**, *62*, 269–84.
95. Ellis-Behnke, R.; *Med Clin of North America.* **2007**,*91* 937–962.
96. Lebedev, M. A.; *Nicolelis MAL. Brain-machine interfaces past, present, future.* Trends Neurosci **2006**, *29*, 536–46.
97. Moxon, K. A.; Hallman, S.; Aslani, A.; Kalkhoran, N.M.; Lelkes, P. I. Bioactive properties of nano-structured porous silicon for enhancing electrode to neuron interfaces. *J Biomater Sci Polym.* **2007**, *18* 1263–1281.
98. Zhao, Y.; Larimer, P.; Pressler, R. T.; Strowbridge, B. W.; Burda, C. Wireless activation of neurons in brain slices using nanostructured semiconductor photoelectrodes. *Angew Chem Int Edn.* **2009**, *48*, 2407–2410.
99. Madou, M. J. Fundamentals of Microfabrication: *The Science of Miniaturisation*, 2nd ed.; CRC Press, Boca Raton, **2002, **752.
100. Borzenko, A. G. Scientifical instrument engineering. **2005**, *15(3)*, 8–24. Analytical devices: Mili- micro- and nanohorizons. (In Russsian) : *15* **2005** *3*, 8–24.)
101. Kim, K.; Lee, J-B. MEMS for drug delivery, Chapter 12. In: Wang, W.; Soper, S. A., eds. *Bio-MEMS: Technologies and Applications*. Boca Raton, FL: CRC Press; **2006**, 325–348.
102. Santini, J. T. Jr.; Richards, A. C.; Scheidt, R. A.; Cima, M. J.; Langer, R. S. Microchip technology in drug delivery. *Ann. Med.*, **2000, **32, 377–379.
103. Santini, J. T., Jr.; Cima, M. J.; Langer, R. A controlled-release microchip. *Nature*, **1999**, *397*, 335–338.
104. Gardner, D. P. Microfabricated nanochannel implantable drug delivery devices: trends, limitations and possibilities. *Exp. Opin. Drug Deliv.*, **2006**, *3*, 479–487.
105. S. Shrivistava; D. Dash Applying nanotechnology to human health. *J. Nanotech* **2009**, 1–14.
106. US Food and Drug Administration. *Nanotechnology page*. Available at: http://www. fda. gov/nanotechnology/. (Accessed February 1, 2010).
107. The Personalized Medicine Coalition. Home page. Available at: http://www.personalized medicinecalitition.org/sciencepolicy/personalmed-101 overview.php. (Accessed January 2, 2010).

108. Amgen, J. Future of biotechnology in healthcare, Chapter 9. In: *An Introduction to Biotechnology*. **2009**, 31–35. Available at: http://www.amgenscholars.eu/web/guest/future-of-biotechnology-in-healthcare (accessed January 2, 2010).

109. Heller. Integrated medical feedback systems for drug delivery. *Am Inst Chem Eng J*, **2005**, *51*, 1054–1066.

110. Brunetti, P.; Federici, M. O.; Benedetti, M. M. The artificial pancreas. *Artif Cells Blood Substit Immobil Biotechnol.* **2003**, *31*, 127–138.

111. Staples, M. Microchips and controlled release drug reservoirs. *WIREs Nanomed. Nanobiotechnol.* **2010**, *2*, 400–417.

112. Genzyme website. *The cost of enzyme replacement therapy.* Available at: http://www.genzyme.com/commitment/patients/costof treatment.asp. (Accessed January 2, **2010**).

113. Shoji, S.; Esashi, M. Microflow Devices and Systems. *J. Micromech. Microeng.* **1994**, *4*, 157–171.

114. Amirouche, F.; Zhou, Y.; Johnson, T. Current micropump technologies and their biomedical applications. *Microsys. Tech.*, **2009**, *15*, 647–667.

INDEX